AF360817

Biocatalysis

Biocatalysis: Chemical Biosynthesis

Editor

Agatha Bastida Codina

MDPI • Basel • Beijing • Wuhan • Barcelona • Belgrade • Manchester • Tokyo • Cluj • Tianjin

Editor
Agatha Bastida Codina
CSIC, Instituto de Química Orgánica General
Spain

Editorial Office
MDPI
St. Alban-Anlage 66
4052 Basel, Switzerland

This is a reprint of articles from the Special Issue published online in the open access journal *Catalysts* (ISSN 2073-4344) (available at: https://www.mdpi.com/journal/catalysts/special_issues/chemical_biosyn).

For citation purposes, cite each article independently as indicated on the article page online and as indicated below:

LastName, A.A.; LastName, B.B.; LastName, C.C. Article Title. *Journal Name* **Year**, *Volume Number*, Page Range.

ISBN 978-3-0365-1230-3 (Hbk)
ISBN 978-3-0365-1231-0 (PDF)

Contents

About the Editor

Agatha Bastida Codina was born in Madrid, Spain, in 1964. She received her degree in Chemistry from the Universidad Autónoma de Madrid. In 1992, she obtained her PhD at the CSIC, working on synthesis of peptides with immobilized enzymes. From 1993 to 1994, she was granted a Postdoctoral Fellowship by the European Community to work at the Research Department of Unilever (United Kingdom) on synthesis of polyol esters with enzymes. In 2007, 2010 and 2018, she worked at Cambridge University and Oxford Brooks University in the UK. Nowadays, Agatha Bastida is an Assistant Research Scientist at the CSIC and she has been working for nearly two decades in a multidisciplinary group (BioGlycoChem Group) with interest in synthetic biology and molecular approaches to problems in the biological sciences using carbohydrates as lead compounds. The research program encompasses the development of computational and experimental methodologies applied to biochemistry of proteins in order to design and produce new active compounds by chemoenzymatic synthesis which medical or industrial value to fight disease.

Preface to "Biocatalysis: Chemical Biosynthesis"

Biocatalysis is a topic on the boundaries of biology and chemistry, which brings together scientists from the life sciences, engineers and computer field. Biocatalysis, cell free and whole cell systems play an increasingly role in a broad range of applications, including food processing, materials, fine chemicals, and medicine. Applied biocatalysis is driven by advances in novel protein engineering tools, economic and environmental pollution. Compared to free biocatalyst, the immobilization techniques improve the stability/activity ration, purification, and the ability to recycle the catalyst in a consequently more efficient process in the industry. Compared to traditional organic synthesis, the use of enzymes provides a way of producing enantiomerical pure compounds, mainly through high chemoselectivity and streoselective properties and very mild reaction conditions, which offer advantages such as minimizing side reactions or not requiring multiple protection/deprotection steps. The advances in directed evolution, molecular cloning, DNA synthesis, bioinformatic tools, high-throughput screening, and process scale-up have opened up a door to a great variety of enzymes as tools in the industry. On the other hand, the use of biosynthetic enzymes can introduce structural diversities that are otherwise inaccessible by organic synthesis. Although biocatalysis has expanded in the industry, a lack of structural and mechanistic knowledge about enzymes has limited the widespread use of enzymes. Then, the combination of in silico methods for engineering physical properties into proteins and structural biology techniques can guide the understanding of biocatalytic mechanisms and the development of new ones in biology and biotechnology.

Agatha Bastida Codina
Editor

 *catalysts*

Editorial

Biocatalysis: Chemical Biosynthesis

Agatha Bastida

Department Bio-organic Chemistry, Institute of General Organic Chemistry (IQOG), The Spanish National Research (CSIC), c/Juan de la Cierva 3, E-28006 Madrid, Spain; agatha.bastida@csic.es

Received: 28 March 2020; Accepted: 31 March 2020; Published: 2 April 2020

Biocatalysis is very appealing for industry because it allows the synthesis of products that are not accessible by chemical synthesis, use of alternative raw materials, lower operating costs, low fixed cost infrastructure and improved eco-efficiency [1]. Chemical biosynthesis has emerged as a valuable tool in organic chemistry that provides straightforward and efficient alternatives to traditional chemical synthesis [2]. Advances in enzyme discovery, high-throughput screening and protein engineering have substantially expanded the available biocatalysts, and consequently, many more synthetic transformations are now possible. The chemo-, regio-, diastereo- and enantio-selectivity of biocatalysts are among the properties that render them superior to chemical catalysts [3,4]. Enzymes are biodegradable and easily replaced through inexpensive and environmentally benign fermentation processes. The minimal environmental and economic burden posed by enzymes is further diminished by the potential to engineer even more active variants than can be found in nature, to optimize expression efficiencies and fermentation yields, and to immobilize and reuse these powerful catalysts. Biocatalysis has been used widely to produce chiral pharmaceutical intermediates to make small molecule drugs [5].

This Special Issue discusses some of the key drivers and scientific developments that are expanding the application of biocatalysis in the pharmaceutical industry through several examples; kinetic characteristics of cofactor-enzyme complex against aldehydes giving important pharmaceutical precursors [6], cyclodextrin glucanotransferase from *Thermoanaerobacter sp* is able to glycosylate a ternary alchols [7,8], bi-enzymatic cascade for the synthesis of optically pure compounds [9], *Sporomusa ovata* used as microbial catalyst to determine whether electron uptake is hydrogen-dependent [10] and esterification of ferulate with octanol catalyzed by lipase in a solvent-free system [11]. In addition, two reviews on biocatalysis as a useful tool in asymmetric synthesis [12] and pseudokinases as suitable drug targets for the treatment of various diseases [13] give an idea about the impact of biocatalysis in drug development.

The research article of Karanam et al. deals with the importance of an (R) specific carbonyl reductase from *Candida parapsilosis* as a biocatalyst in the reduction of ketones and keto-esters of aliphatic and aromatic aldehydes [6]. García-Arellano et al. investigated the enzymatic synthesis, using a cyclodextrin glycosyltransferase from *Thermoanaerobacter sp*, of novel alkyl glycoside derivatives which are used as non-ionic surface-active agents in food, cosmetic and detergent industries [7]. Chiang et al. studied the enzymatic synthesis of new isoflavone glucosides, 8-OHDe-7/8-O-β-glucoside, increasing its solubility and stability to be used in pharmaceutical and cosmeceutical applications [8]. The other paper in this Special Issue focuses on a particular application in the use of enzymatic cascades in one pot for the synthesis of active cyanohydrins [9]. Leemans et al. attempted to couple the hydrocynation of 4-anisaldehyde with the acetylation of the formed mandelonitrile catalyzed by *Manihot esculenta* hydroxynitrile lyase (an S-selective enzyme) and *Candida antartica* lipase A. Tremblay et al. used *Sporomus ovata* as the microbial catalyst to determine whether electron uptake is hydrogen dependent. Hydrogen was detected at −500 mV versus Ag/AgCl by a micro-sensor in proximity to the cathode in a sterile fresh medium. As well, nickel and cobalt were detected at the cathode surface after exposure to the medium increasing the presence of H_2. So, at least a part of the electrons coming from the electrode are transferred to the bacteria via hydrogen during microbial electrolysis [10]. Huang et al. described the enzymatic esterification of ferulic acid with octanol producing octyl ferulate in a high-temperature solvent-free system with an environmentally friendly method. Due to the limited solubility of ferulic acid in a lipophilic medium, the ester form of ferulic acid has better antioxidant and ultraviolet-absorbing activity [11].

We have two review papers in this Special Issue, one by Domínguez de María et al., presenting an excellent overview of recent progress on the assessment of granted patents about biocatalysis as a useful tool in asymmetric synthesis [12]. A vast area of enzyme is used and is performed by industry or academic groups. Around 200 granted patents have been identified, covering all enzyme types. The inventive pattern focuses on the protection of novel protein sequences, as well as on new substrates. In some other cases, combined processes, multi-step enzymatic reactions, as well as process conditions are the innovative basis. The most relevant found in this review was the protection of novel protein sequences for well-known reactions and the use of novel substrates for useful synthetic reactions. Biocatalysis is taking significant steps in chemical industries, being seriously considered as a powerful alternative for combining sustainable chemistry with high efficiency and selectivity.

A second review, by Jonathan Lees et al., deals with a class of pseudokinases enzymes and their potential as suitable drug targets for the treatment of various diseases, including metabolic, neurological, autoimmune, and cell proliferation disorders [13]. Pseudokinases are a subset of the protein kinase superfamily that present inactivating mutations in critical catalytic motifs, but signal primarily through non-catalytic mechanisms. These proteins play central roles in the cell, and the loss of their regulation can lead to a variety of diseases. Indeed, pseudokinase malfunctions occur in diverse diseases and represent a new therapeutic window for the development of innovative therapeutic approaches. Several clinically approved kinase inhibitors have been shown to influence the non-catalytic functions of active kinases, providing the expectation that similar properties in pseudokinases could be pharmacologically regulated. Recent advances in drug discovery, assisted by the use of artificial intelligence approaches, may pave the way for the rapid identification of potent pseudokinase inhibitors.

Thanks to all the authors for their valuable contributions and the editorial team of *Catalysts* for their kind support, without them, this Special Issue would not have been possible.

Conflicts of Interest: The authors declare no conflicts of interest.

References

1. Devine, P.N.; Howard, R.M.; Kumar, R.; Thompson, M.P.; Truppo, M.D.; Turner, N.J. Extending the application of biocatalysis to meet the challenges of drug development. *Nat. Rev. Chem.* **2018**, *2*, 409–421. [CrossRef]
2. Chouhan, S.; Sharma, K.; Zha, J.; Guleria, S.; Koffas Mattheos, A.G. Recent Advances in the Recombinant Biosynthesis of Polyphenols. *Front Microbiol.* **2017**, *8*, 2259. [CrossRef] [PubMed]
3. Martinez-Montero, L.; Schrittwieser, J.H.; Kroutil, W. Regioselective Biocatalytic Transformations Employing Transaminases and Tyrosine Phenol Lyases. *Top. Catal.* **2019**, *62*, 1208. [CrossRef]
4. Li, G.; Wang, J.B.; Reetz, M.T. Biocatalysts for the pharmaceutical industry created by structure-guided directed evolution of stereoselective enzymes. *Bioorg. Med. Chem.* **2017**, *26*, 1241–1251. [CrossRef] [PubMed]
5. Su, H.; Zhang, H.; Ang, E.L.; Zhao, H. Biocatalysis for the synthesis of pharmaceuticals and pharmaceutical intermediates. *Bioorg. Med. Chem.* **2018**, *26*, 1275–1284. [CrossRef]
6. Karanam, V.K.; Chaudhury, D.; Chadha, A. Understanding (R) Specific Carbonyl Reductase from *Candida parapsilosis* ATCC 7330 [CpCR]: Substrate Scope, Kinetic Studies and the Role of Zinc. *Catalysts* **2019**, *9*, 702. [CrossRef]
7. Garcia-Arellano, H.; Gonzalez-Alfonso, J.L.; Ubilla, C.; Comelles, F.; Alcalde, M.; Bernabé, M.; Parra, J.-L.; Ballesteros, A.O.; Plou, F.J. Production and Surfactant Properties of *Tert*-Butyl-D-Glucopyranosides Catalyzed by Cyclodextrin Glucanotransferase. *Catalysts* **2019**, *9*, 575. [CrossRef]

8. Chiang, C.-M.; Wang, T.-Y.; Yang, S.-Y.; Wu, J.-Y.; Chang, T.-S. Production of New Isoflavone Glucosides from Glycosylation of 8-Hydroxydaidzein by Glycosyltransferase from *Bacillus subtilis* ATCC 6633. *Catalysts* **2018**, *8*, 387. [CrossRef]

9. Leemans, L.; Langen, L.V.; Hollmann, F.; Schallmey, A. Bi-enzymatic Cascade for the Synthesis of an Optically Active O-benzoyl Cyanohydrin. *Catalysts* **2019**, *9*, 522. [CrossRef]

10. Tremblay, P.-L.; Faraghiparapari, N.; Zhang, T. Accelerated H_2 Evolution during Microbial Electrosynthesis with *Sporomusa ovata*. *Catalysts* **2019**, *9*, 166. [CrossRef]

11. Huang, S.-M.; Wu, P.-Y.; Chen, J.-H.; Kuo, C.-H.; Shieh, C.-J. Developing a High-Temperature Solvent-Free System for Efficient Biocatalysis of Octyl Ferulate. *Catalysts* **2018**, *8*, 338. [CrossRef]

12. Gonzalo de Gonzalo, P.D.; Alcántara, A.R. Biocatalysis as Useful Tool in Asymmetric Synthesis: An Assessment of Recently Granted Patents (2014–2019). *Catalysts* **2019**, *9*, 802. [CrossRef]

13. Jonathan Lees, A.T.; Santana, A.G.; Bolanos-Garcia, V.M.; Bastida, A. Pseudokinases: From Allosteric Regulation of Catalytic Domains and the Formation of Macromolecular Assemblies to Emerging Drug Targets. *Catalysts* **2019**, *9*, 778. [CrossRef]

Review

Biocatalysis as Useful Tool in Asymmetric Synthesis: An Assessment of Recently Granted Patents (2014–2019)

Pablo Domínguez de María [1,*], **Gonzalo de Gonzalo** [2,*] and **Andrés R. Alcántara** [3,*]

[1]   Sustainable Momentum, SL. Av. Ansite 3, 4–6, E-35011 Las Palmas Gran Canaria, Spain
[2]   Organic Chemistry Departament, University of Sevilla, c/ Profesor García González 2, 41012 Sevilla, Spain
[3]   Department of Chemistry in Pharmaceutical Sciences, Section of Organic Chemistry (Pharmaceutical Chemistry), Faculty of Pharmacy, Complutense University of Madrid, Plaza de Ramón y Cajal, s/n., E-28040 Madrid, Spain
*   Correspondence: dominguez@sustainable-momentum.net (P.D.d.M.); gdegonzalo@us.es (G.d.G.); andalcan@ucm.es (A.R.A.); Tel.: +34-60-956-52-37 (P.D.d.M.); +34-95-455-99-97 (G.d.G.); +34-91-394-18-20 (A.R.A.)

Received: 6 September 2019; Accepted: 22 September 2019; Published: 25 September 2019

Abstract: The broad interdisciplinary nature of biocatalysis fosters innovation, as different technical fields are interconnected and synergized. A way to depict that innovation is by conducting a survey on patent activities. This paper analyses the intellectual property activities of the last five years (2014–2019) with a specific focus on biocatalysis applied to asymmetric synthesis. Furthermore, to reflect the inventive and innovative steps, only patents that were granted during that period are considered. Patent searches using several keywords (e.g., enzyme names) have been conducted by using several patent engine servers (e.g., Espacenet, SciFinder, Google Patents), with focus on granted patents during the period 2014–2019. Around 200 granted patents have been identified, covering all enzyme types. The inventive pattern focuses on the protection of novel protein sequences, as well as on new substrates. In some other cases, combined processes, multi-step enzymatic reactions, as well as process conditions are the innovative basis. Both industries and academic groups are active in patenting. As a conclusion of this survey, we can assert that biocatalysis is increasingly recognized as a useful tool for asymmetric synthesis and being considered as an innovative option to build IP and protect synthetic routes.

Keywords: biocatalysis; asymmetric synthesis; patents; lipases; oxidoreductases; lyases; transaminases

---

1. Motivation. The Role of Granted Patents as Key Actors in Innovation

Biocatalysis is currently recognized as a mature technology for asymmetric synthesis, based on the excellent selectivity (enantio-, chemo, and regio-) that enzymes often display in many chemical processes. The field has experienced a tremendous development when emerging Molecular Biology techniques, used to produce and design enzymes, have been synergized with both process-development concepts (batch to continuous reactors) and medium engineering strategies (non-aqueous media, immobilization, etc.). As a result, biocatalysis can fulfil, in many cases, the demands of modern organic synthesis, which must align a high selectivity and efficiency with sustainability and tight economic figures. As a result, an increasing number of industrial processes have been successfully established, comprising many diverse examples using free enzymes, whole cells, systems biocatalysis, multi-step processes, etc. [1–19].

The abovementioned broad interdisciplinary profile makes biocatalysis in a fertile landscape for innovation. It is common that researchers of different backgrounds (chemistry, biology, engineering, etc.)

cooperate closely in setting up new enzymatic reactors [20–23]. Overall, each biocatalytic industrial reaction contains highly evolved parts, which are not standing alone, but that do form a robust process when all of them are combined smartly. Some key concepts for industrial biocatalysis may be: i) designed enzyme variants with improved selectivity, high resistance to chemicals, thermostability, etc. [24–30]; ii) bio-based (non-conventional) solvents used as co-solvents in water, or as biphasic systems [31–38]; neoteric solvents, such as deep eutectic solvents (DES) [39–44]; iii) whole-cells [45–48], immobilization [49–56], continuous processes, etc. [12,23,57,58]; and iv) metrics [59–62], high productivities to assure an industrially-sound reaction.

To obtain deeper insights into these expected innovations, this paper discloses a patent survey related to biocatalysis applied to asymmetric synthesis. Although this genre of studies is (surprisingly) rather scarce in the open literature [8,63,64], we believe that the approach may nicely reflect the modern trends in the field, provided by academic and industrial groups, and comprising different enzymes and chemical processes. Importantly, to further refine the innovation assessment, the focus of the study is put solely on those patents that have been granted over the last five years (2014–June 2019). Remarkably, granted patents may provide further proofs on the necessary steps of novelty, innovation, and resolution of industrial problems, as demanded for intellectual property (IP) rights. To simplify the reading of the article, the granted patents are discussed by enzyme types, covering several relevant case studies. Moreover, the focus of the article is strictly scientific, and no information regarding the countries in which the invention is granted, nor whether or not patent fees are being satisfied, is provided. It must be noted, however, that readers can easily retrieve that information from the patent servers, by simply tracking the patent number provided in the references. To follow a consistency with publication rules, the years denoted in the references are the publication years; in all cases, the effective dates of those granted patents mentioned in this article are included between 2014–2019, and it can be retrieved from patent servers as well. According to the survey, around 200 patents related to biocatalysis have been successfully granted within the last five years, as shown in Figure 1.

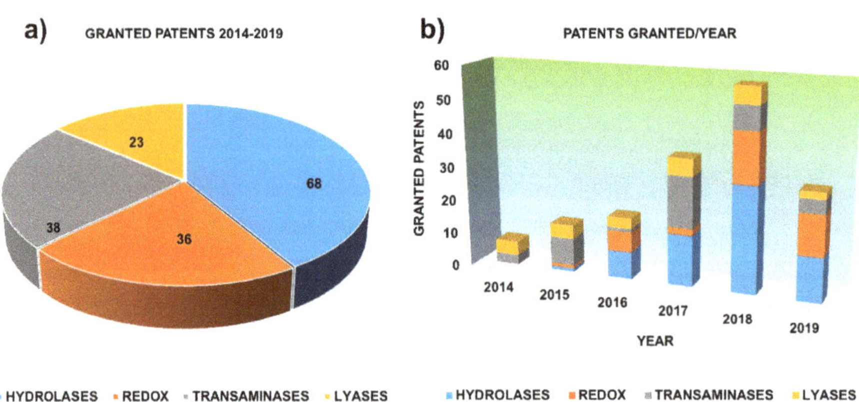

Figure 1. Patents granted in 2014–2019 employing the most usual types of enzymes, according to several patent engine servers (e.g., Espacenet, SciFinder, Google Patents).

Notably, examples for all enzyme types have been found, for asymmetric synthesis purposes, and the main strategies to achieve patentability are either the protection of known enzymes and variants with different sequences (covering up to some homology), or the use of novel substrates for the intended reactions, or a combination of both approaches. Apart from industrial activities – pharmaceutical and fine chemistry firms – several academic groups have proven to be active in IP activities as well.

2. Granted Patents Related to Hydrolases in Asymmetric Synthesis

2.1. Lipases

Hydrolases (Enzyme Commission number EC 3,) undeniably represent the most valuable type of enzymes in applied biocatalysis, due to their unique properties, such as broad substrate specificity, commercial availability, lack of need to use any cofactor, and high capability to work at high substrate concentrations [65–73]. Inside this group of enzymes, lipases (EC 3.1.1.3) can be considered the most useful, according to the ample number of processes in which they have been employed at lab [74–77] and industrial scale [78–82], mainly because they can perform with notable precision (chemo-, regio- and/or stereoselectivity) not only in their natural environment (aqueous solutions) but also in non-conventional media [66,75,79,82–85], therefore being catalogued as natural promiscuous enzymes [86]. Although the main industrial uses of lipases are in the detergent and food and beverage areas, there are plenty of examples in which these enzymes are employed in asymmetric synthesis [63]; in this sense, examples taken from granted patents in the last five years are hereunder presented.

2.1.1. Stereoselective Hydrolysis Catalyzed by Lipases

Thus, regarding the employ of lipases in the preparation of chiral building blocks for the synthesis of active pharmaceutical ingredients (APIs), some recent patents describe different stereoselective hydrolytic procedures taking advantage of the enzymatic capability of discriminating between enantiomers. For instance, a patent from Zhejiang Changming Pharmaceutical Co., Ltd., [87], granted in 2018, describes the preparation of the anti-epileptic drug levetiracetam (Keppra®, Elepsia®) starting from racemic methyl 2-bromo(chloro)butanoate, using Novozym 435, a commercially available preparation of lipase B from *Candida antarctica* immobilized on Lewatit VP OC 1600, a macroporous acrylic polymer resin [88]. The key step, a kinetic resolution of a racemic 2-haloester **1**, leads to the (*S*)-acid and the desired ester (*R*)-**2**, which is extracted using ethyl acetate and converted into levetiracetam via transformation of the (*R*)-ester into the α-aminoamide (*S*)-**3** and a further cyclization with 4-chlorobutanoyl chloride, as shown in Scheme 1.

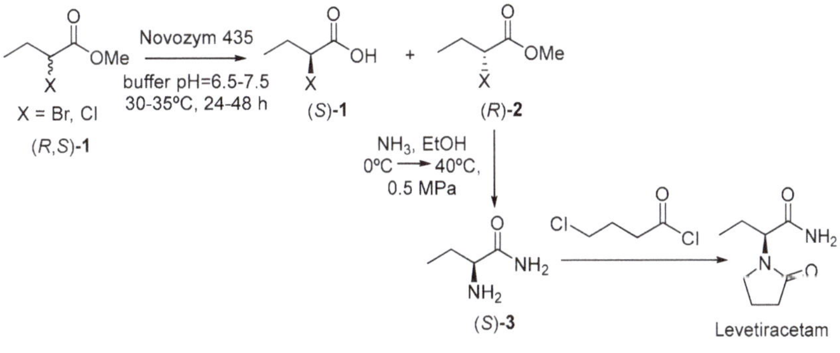

Scheme 1. Chemoenzymatic preparation of levetiracetam starting from a racemic 2-haloester.

In another example, developed by Daiichi Sankyo Co. Ltd., and granted in 2019 [89], the kinetic resolution of (*R*,*S*)-1-(((((1,1,1,3,3,3-hexafluoropropan-2-yl)oxy)carbonyl)oxy)ethyl isobutyrate ((*R*,*S*)-**4**) was carried out by a lipase-catalyzed hydrolysis in a buffer/organic solvent biphasic system to discard the corresponding alcohol (*S*)-**5** from the desired ester (*R*)-**4** (Scheme 2).

In order to catalyze this enzymatic hydrolysis, different commercial lipases (from *C. antarctica*, *Candida rugosa* and *Thermomyces lanuginosus*), organic cosolvents (acetonitrile, ethers, saturated and aromatic hydrocarbons, alcohols) and reaction conditions (temperature from 0 °C to 80 °C, reaction times ranging from 1 h to 120 h) were tested; the best results were obtained with Chirazyme L2, C4 (a commercial preparation of *C. antarctica* lipase similar to Novozym 435) and buffer pH=7/MTBE

(5/1, V/V) at 20 °C, with reported conversions of around 40% and ee values higher than 99%. Subsequently, (*R*)-**4** was used to furnish a carbamate **7** upon reaction with the enantiopure amide **6** (as a sulphonic salt). This carbamate **7** is a promising prodrug useful as inhibitor of carboxypeptidase U (also known as activated thrombin-activatable fibrinolysis inhibitor, TAFIa), for treating thromboembolic disorders [90–92].

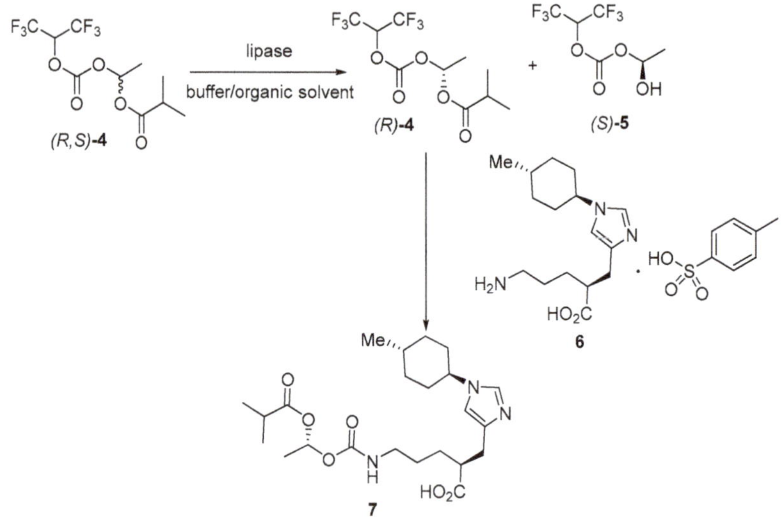

Scheme 2. Synthesis of prodrug **7** possessing TAFIa inhibitory activity.

Another enzymatic hydrolysis useful for synthesizing drug precursors was patented by Beijing University of Chemical Technology [93] and granted in 2017; the stereoselective hydrolysis of a racemic mixture of *N*-protected *cis*-4-amino-cyclopent-2-en-1-ol (*cis*-**8**, 20 g/L, Scheme 3) was carried out using lipase B from *C. antarctica* at pH = 5 and 60 °C in 4 h, to furnish the desired alcohol (1*S*,4*R*)-**9** (conversion 45%, ee 99%). This enantiopure alcohol is a building block for the preparation of ticagrelor (Brilinta®, Brilique®, Possia®) a platelet aggregation inhibitor originally produced by AstraZeneca [94,95].

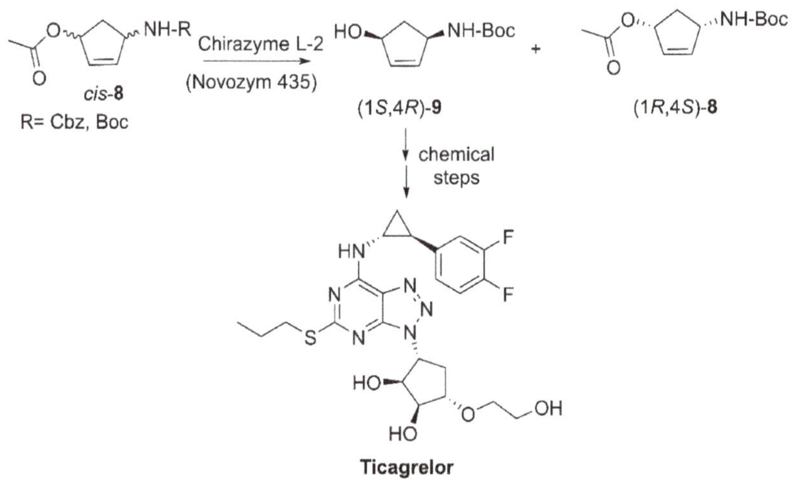

Scheme 3. Synthesis of an enantiopure precursor of ticagrelor.

A very interesting process, patented by and recently granted in 2019 to AbbVie Inc. describes preparation of an intermediate ((1*R*, 2*R*)-**12**, Scheme 4) in the synthesis of viral protease inhibitors [96]. In this case, a double enzymatic procedure is followed starting from (±)-dipropyl 2-(difluoromethyl)cyclopropane-1,1-dicarboxylate (±)-**10**, and employing primarily a lipase (from *T. lanuginosus* or *Rhizomucor miehei*) in a biphasic medium (phosphate buffer with 10% DMF, 30 °C, 95 h) to separate the undesired monoacid (1*R*, 2*S*)-**11** (extracted from the reaction media using MTBE) from the chiral diester (*R*)-**10** (64.4% yield, 97.4% ee). In a second step, an esterase from *Bacillus subtilis* cloned and expressed in cells from *Escherichia coli* was employed for the desymmetrization of the stereocentre at C1 of (*R*)-**10** to yield (1*S*, 2*R*)-**10** (84.2% yield, 99.4% ee) in an aqueous solution at 30 °C in 68 h.

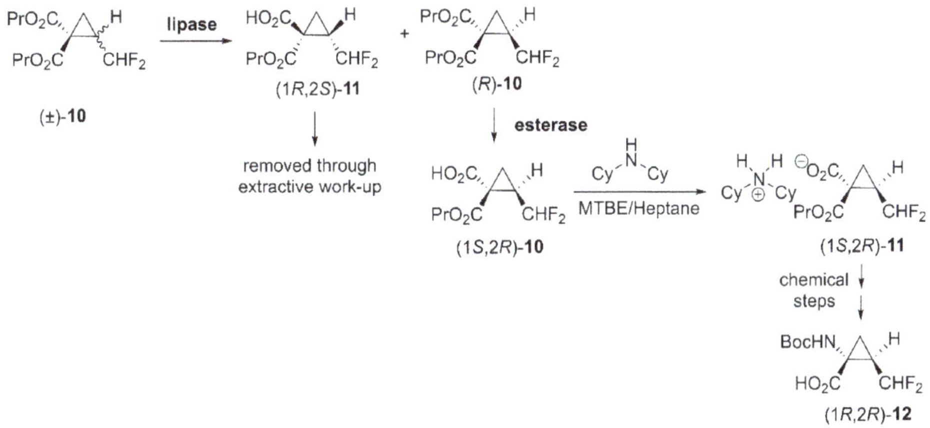

Scheme 4. Chemoenzymatic synthesis of an enantiopure precursor of a viral protease inhibitor.

All the patents mentioned up to this point share the common feature of placing the enzymatic asymmetric hydrolysis in an initial stage of the synthetic scheme leading to the desired API. More infrequently, the enzymatic biotransformation is applied once the structure of the API is almost complete. For instance, a patent from Tiantai Yisheng Biochemical Technology Co., Ltd., (granted in 2019) describes the enantioselective hydrolysis of racemic esters of valsartan (Diovan®, an angiotensin II receptor blocker (ARBs or sartans), a drug which modulates the renin–angiotensin system, used in the treatment of hypertension, diabetic nephropathy and congestive heart failure [97]) catalyzed by lipases [98], as depicted in Scheme 5.

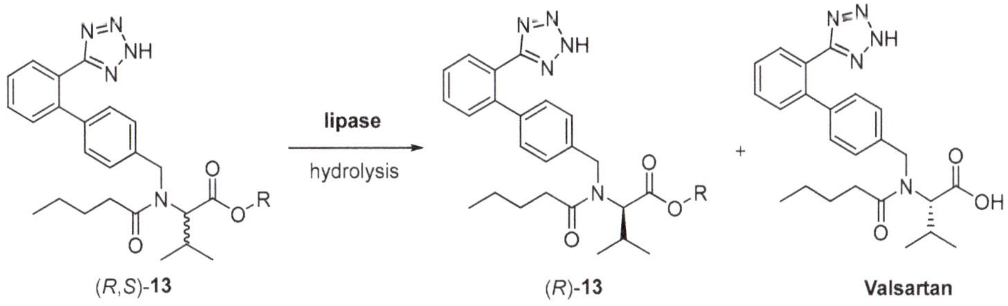

Scheme 5. Enzymatic resolution of racemic esters of valsartan.

Thus, different commercial immobilized lipases (Lipozyme TL IM (*T. lanuginosus*), Lipozyme RM IM (*R. miehei*), Novozym 435 (lipase B from *C. antarctica*) from Novozymes; lipases PS IM (*Burkholderia cepacia*), Lipase AK (*Pseudomonas fluorescens*) and Lipase AS (*Aspergillus niger*), from Amano) were used for the hydrolysis of racemic esters ((*R*,*S*)-**13**, R = ethyl, methyl or isopropyl) in a buffer solution (pH varying between 4.0–9.0) with different amounts of organic cosolvents (acetone, *iso*propanol or ethyl acetate), rendering the desired valsartan ((*S*) absolute configuration) in good yields (up to 90–95% after work-up) and enantiopurity (up to 99.5%).

2.1.2. Regio- and Chemoselective Hydrolysis Catalyzed by Lipases

Lipases can be also used for the regioselective hydrolysis of a specific ester group without altering another similar functional groups placed in another part of the molecule, under very mild reaction conditions. In a recent patent from Harbin University of Commerce, granted in 2019 [99], a lipase (from *Candida sp.* among others) catalyzing the monohydrolysis of nifedipine (dimethyl 2,6-dimethyl-4-(2-nitrophenyl)-1,4-dihydropyridine-3,5-dicarboxylate (Scheme 6), a calcium channel blocker [100]) is reported to furnish the corresponding monoacid **14**; this reaction is carried out in a mixed system of aq. soln./org. solvent (such as isooctane, hexane, cyclohexane or toluene), and reacting under stirring at 30–40 °C for 3–5 days. The monoacid (23% yield, no ee reported) is the key step for synthesizing other 1,4-dihydropyridine structures possessing two different alkoxycarbonyl groups in positions 3 and 5 (such as aranidipine, azeldinipide, barnidipine, etc.), which are more selective and efficient drugs [101].

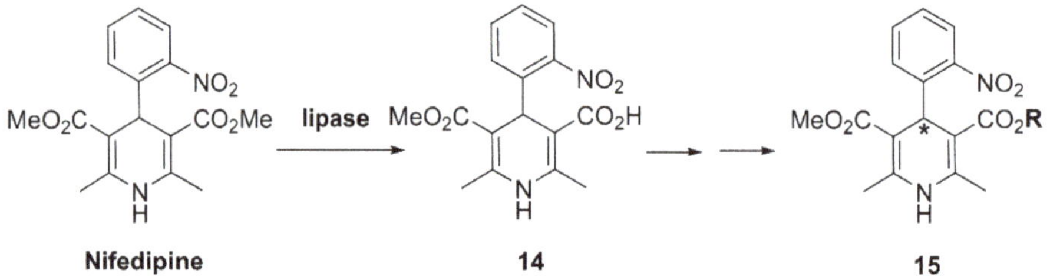

Scheme 6. Enzymatic mono-hydrolysis of nifedipine.

The possibility of performing chemoselective reactions is another attractive property of lipases. In this sense, a patent from AstaTech Chengdu Biopharmaceutical Co., Ltd. granted in 2018 [102], reports the use of several lipases (Novozym 435 and lipase from pig pancreas leading to the best results) and esterases for the hydrolysis of different alkyl esters of *N*-Boc-(1*R*,3*S*,4*S*)-2-azabicyclo[2.2.1]heptane-3-carboxylate (**16**, Scheme 7), to yield acid **17** without removing the carbamate protecting group. In this process, essential for the preparation of ledispavir (Harvoni®, Gilead, in combination with sofosbuvir (see Scheme 14) for treatment of hepatitis C [103], the enzymatic hydrolysis allows the preparation of **17** in good yields (up to 77%) and high d.e. values (up to 99.7%), therefore avoiding the main problem of the chemical basic hydrolysis (LiOH in THF/H$_2$O), a partial racemization of the stereogenic center at Cα of the alkoxycarbonyl group.

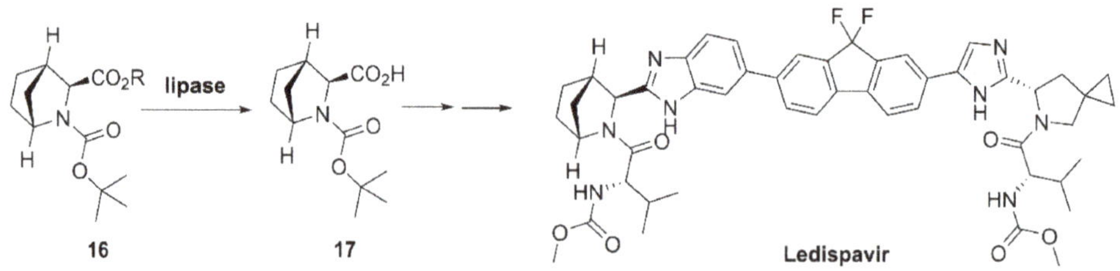

Scheme 7. Enzymatic preparation of a ledispavir intermediate.

2.1.3. Stereoselective Acyl-Transfer Catalyzed by Lipases

As stated above [66,75,79,82–85], lipases are perfectly capable of working in non-aqueous reaction media, therefore catalyzing acyl transfer processes (esterifications, transesterifications, amidations, etc) with excellent precision. Hence, in a patent from Ningbo Xinkai Biotechnology Co., Ltd., granted in 2017 [104], the use of Nozozym 435 for the stereoselective monoacylation of prochiral 2-(2-(2,4-difluorophenyl)allyl)propane-1,3-diol **18** (Scheme 8) with isopropanoic anhydride in a mixture of NaHCO$_3$/toluene at 15 °C during 24 h is reported to furnish monoester **19** (75% yield, ee not reported), a chiral building block for the preparation of antifungic posaconazole (Noxafil®, Posanol®). This biocatalytic monoacylation is not the real innovation of the patent, because it was described previously for this same enzyme using vinyl acetate in acetonitrile [105]; in this case, the synthesis of starting diol **18** is claimed to improve the previously reported procedures [106–108].

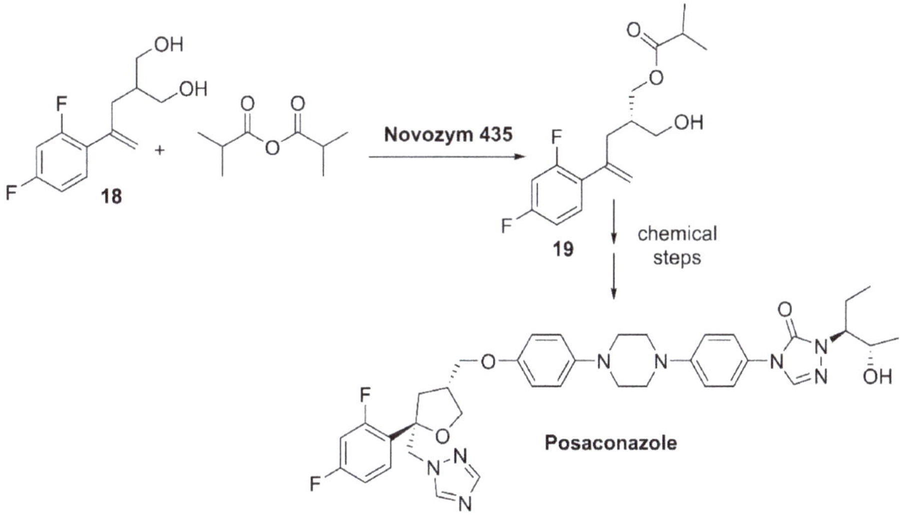

Scheme 8. Chemoenzymatic preparation of an intermediate in the synthesis of posaconazole.

The use of biotransformations in the synthesis of statins, inhibitors of 3-hydroxy-3-methylglutaryl coenzyme A (HMG-CoA) reductase, widely prescribed for the pharmacological treatment of hypercholesterolemia and dyslipidaemia, has been recently reviewed [109,110] In this context, we will comment some recently granted patents using different enzymes.

For instance, Shandong Qidu Pharmaceutical Co., Ltd. patented a transesterification protocol, granted in 2016 [111] and schematized in Scheme 9 for the preparation of pitavastatin (Livalo®, Livazo®, among others trade names), a member of the superstatin family [109].

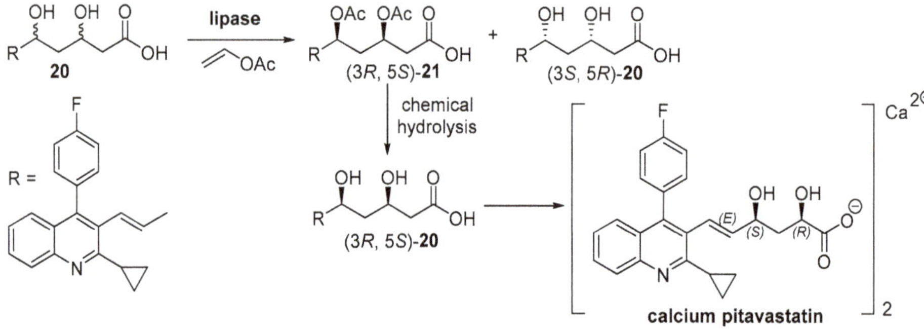

Scheme 9. Preparation of Pitavastatin through an enzymatic transesterification.

In this process, the biocatalyst (commercial lipases from different origins, the best results being obtained with lipase from *B. cepacia*) is able to discriminate between the stereoisomers of starting diol **20** (it is not specified in the patent if **20** is a mixture of the four possible stereoisomers, although according to the reported results it must be a racemic mixture of *syn* enantiomers). The acylation is carried out in a biphasic medium phosphate buffer/organic solvent (dichloromethane, ethyl acetate, tetrahydrofuran, methyl *tert*-butyl ether, cyclohexane, toluene or xylene) in different ratios (from $\frac{1}{2}$ to 1/5, V/V) at temperatures ranging from 0 to 40 °C, and different reaction times (up to 72 h). Reaction yields varied between 43.1 and 47.3%, with excellent enantiopurity of the diacetate (3R, 5S)-**21** (97–99%), which is separated (by ethanol precipitation) from the non-converted diol (3S, 5R)-**20**, and subsequently hydrolyzed chemically (EtOH/NaOH) and stirred with a CaCl$_2$ solution to finally furnish the calcium salt of pitavastatin.

Treprostinil (Remodulin®, Orenitram®, Tyvaso®, Scheme 10), a synthetic analog of prostacyclin (PGI2), is a vasodilator used for the treatment of pulmonary arterial hypertension [112], of which synthesis is complex and time consuming [113]. In a patent from Everlight Chemical Industrial Corporation granted in 2016, a stereo- and regioselective acylation of diol **21** with vinyl acetate, performed in *n*-hexane at 22 °C with lipase AK from Amano, leads to the regioselective acylation of the hydroxyl group in the cyclic moiety of **21** without acylating the one on the exocyclic chain, and leading to the desired stereochemistry of **22** (no yields or ee reported).

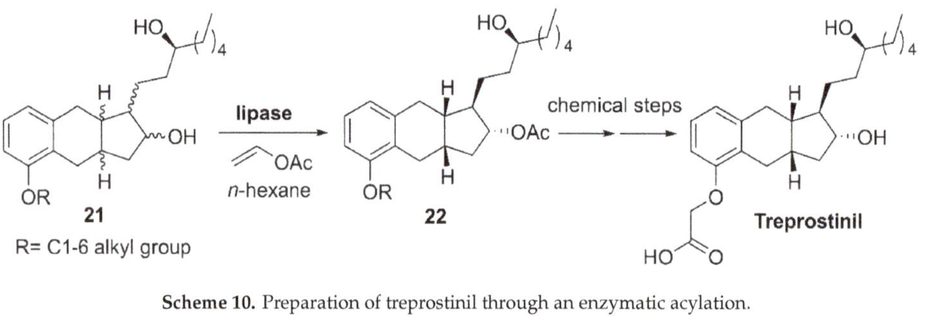

Scheme 10. Preparation of treprostinil through an enzymatic acylation.

In another patent from Wuxi Fortune Pharmaceutical Co., Ltd., granted in 2018 [114], the gram-scale kinetic resolution (Scheme 11) of racemic hidroxynitrile (*R*,*S*)-**24**, obtained from racemic epyclorhydrine (*R*,*S*)-**23**, was reported using an enzymatic transesterification with perchlorophenyl acetate and lipase B from *C. antarctica*, supported on an acid resin (Novozym 435?), to obtain acetate (*R*)-**25**, which hydrolysis leads to L-carnitine, an essential co-factor in the metabolism of lipids involved in the generation of cellular energy [115,116].

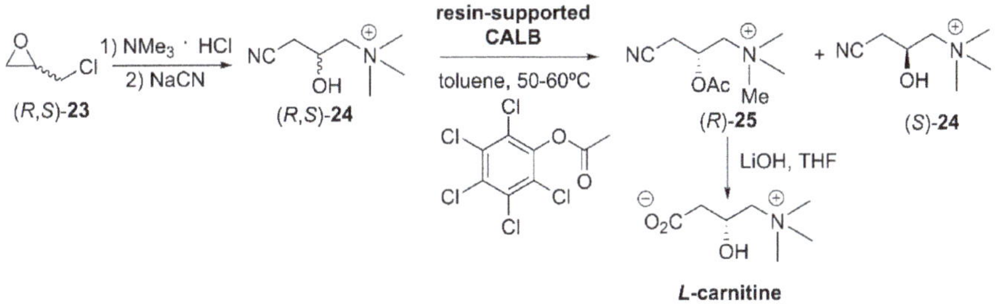

Scheme 11. Chemoenzymatic preparation of L-carnitine.

The resolution of chiral amines is a very attractive research area, because of the plethora of applications described for such enantiopure compounds [117,118], which can be obtained by using oxidoreductases [119] (see Section 3.2), transaminases [120] (Section 4), and lipases [121,122], as we will comment here. In fact, different companies have been actively working in this area during the last years. For instance, in patent granted in 2018 [123], a methodology for preparing enantiopure (R)-1,2,3,4-tetrahydronaphthalen-2-amine (2-aminotetralin, 2-AT, (R)-27, Scheme 12a), a rigid analogue of phenylisobutylamine capable to inhibit the reuptake of serotonin and norepinephrine [124] was reported. In this case, the reaction was carried out in a 2 L autoclave filled with 1 L of toluene, 117.6 g of (R,S)-27 (obtained via reductive amination of starting ketone 26), 172 g of (R)-O-acetyl mandelic acid, commercial Novozym 435 and H$_2$/nickel catalyst (KT-02) to transform the kinetic resolution into a dynamic kinetic resolution by racemizing the unreacted (S)-27. After 24 h, 144.2 g of (R)-28 (95% yield, 99% ee) were isolated and subsequently hydrolyzed to finally yield (R)-28 with excellent yield (92%) and enantiopurity (99%).

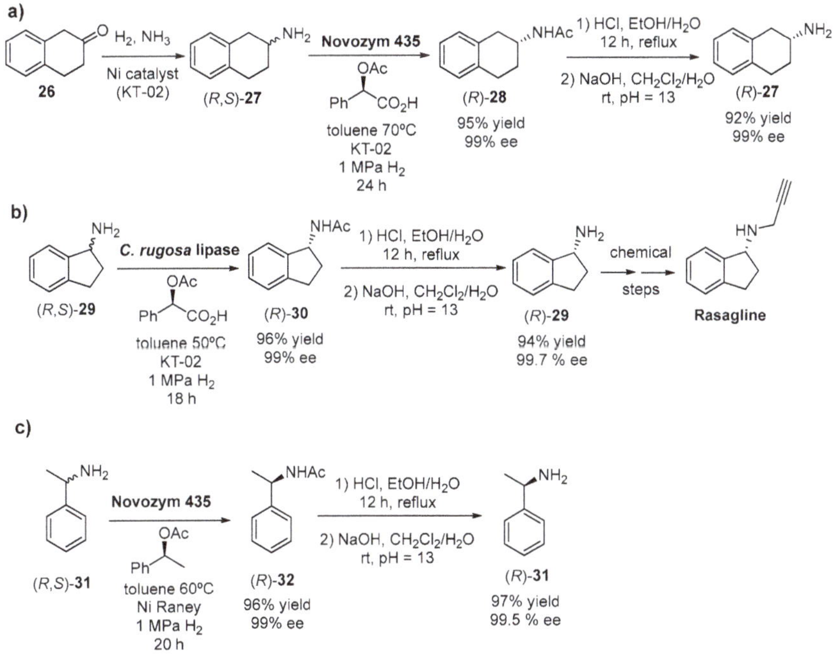

Scheme 12. Lipase-catalyzed preparation of enantiopure amines.

In another very similar example shown in Scheme 12b, the dynamic kinetic resolution of (*R,S*)-**29** was reported in a patent [125] granted in 2018 for the preparation of an enantiopure intermediate in the synthesis of rasagline (Azilect®), an irreversible inhibitor of monoamine oxidase-B, used as a single therapy in the symptomatic treatment in early stages of Parkinson's disease or as an additional therapy in more severe cases [126,127]. In this case, the lipase used is different, but the final results (0.5 L reaction volume) are also excellent in terms of both yield and enantiopurity. Finally, the preparation of optically pure (*R*)-1-phenylethan-1-amine ((*R*)-**31**, Scheme 12c) is described following a similar methodology in another patent [128] granted in 2017.

2.1.4. Site-selective Acyl-Transfer Catalyzed by Lipases

The lipase capability for catalyzing site-selective acylations/deacylations has been also exploited in recently granted patents. For instance, a HC-Pharma AG patent granted in 2018 [129] describes the mono-deacetylation of **33** (Scheme 13), a precursor in the preparation of sofosbuvir (Sovaldi®, Gilead), a direct acting antiviral medication used as part of combination therapy to treat chronic hepatitis C [130].

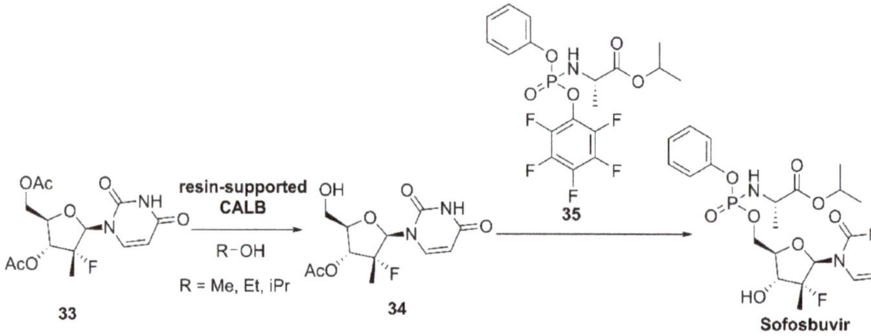

Scheme 13. Chemoenzymatic synthesis of sofosbuvir.

The deacetylation is carried out using 500 g of **33** in 5 L of an alcoholic medium (isopropanol, methoxypropanol or ethanol/water) *via* a single transesterification at 60° during 48 h of the acetoxy group at position 5′ using 100 g of a resin-supported lipase B from *C. antarctica* (Novozym 435?) as the catalyst, leading to quantitative yields (between 99.2 and 99.6%) after a simple purification procedure (cooling to 25 °C, filtering out the lipase, solvent distillation to a residual volume of 900 mL and precipitation at 0 °C).

The site-selective monoacylation of the sugar moiety of useful glycosides has been also patented. For instance, acylation of polydatin (3,4,5-trihydroxystilbene-3-β-D-glucopyranoside), a pharmacological compound isolated from the root and rhizome of the traditional Chinese medicinal plant *Polygonum cuspidatum*, widely applied in anti-inflammatory, anti-oxidant and in anti-angiogenesis chemotherapy [131,132], has been reported in a patent from Huaiyin Institute of Technology, granted in 2018 [133]. This process, depicted in Scheme 14a, uses immobilized lipase from *C. antarctica* (Novozym 435?), amongst others, in 2-MeTHF, a biogenic solvent considered a renewable alternative to THF and other organic solvents [33,134], to furnish almost quantitative yields of **36**, the monoacyl derivative at the OH group at C6 of the glucopyranoside moiety, a useful prodrug of polydatin.

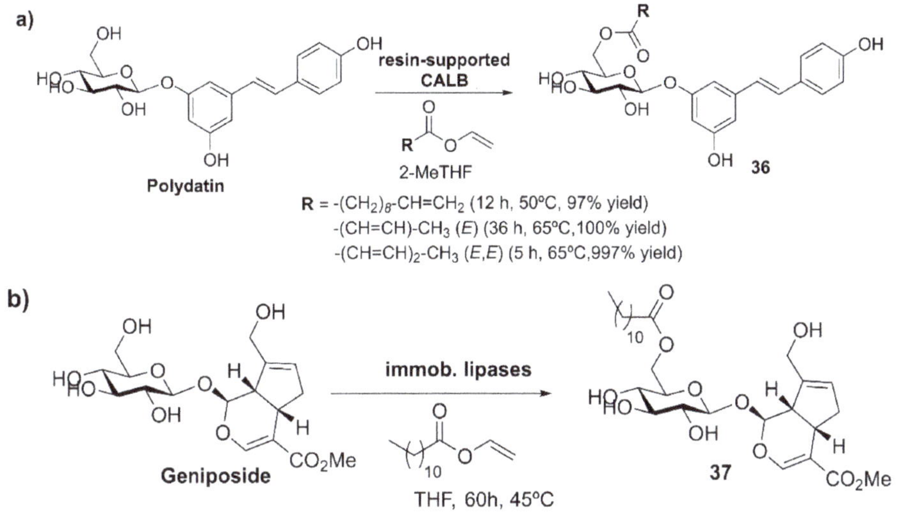

Scheme 14. (**a**) *Cont.* (**b**) Regioselective monoacylation of glycosides.

Geniposide is an important component of *Gardenia jasminoides* Ellis, a plant from Yinchenhao Tang, useful in the prevention and therapy of hepatic injury (HI) [135]. The site-selective monoacylation of geniposide (Scheme 14b) to furnish the more active and stable C-6′-lauroyl monoester has been reported in a patent from Nanjing Tech University, granted in 2019 [136], using different lipases (*Rh. miehei*, lipases A and B from *C. Antarctica* (Novozym 435?)) adsorbed on silica gel (easily separated by filtering after reaction) in THF at 45 °C (yields up to 78%).

In another example, the monoacetylation of (*Z*)-5-ethylidene-8-hydroxy-7-((3-hydroxypyridin-4-yl) methyl)-3,4,5,6,7,8-hexahydro-1*H*-pyrano[3,4-c]pyridin-1-one (**38**, Scheme 15), derived from the iridiod-type aglycon of multi-active swertiamarine [137–139], is reported in a patent from Jiangxi Science & Technology Normal University granted in 2018 [140]. In this patent, compound **38** is obtained by a controlled fermentation of *Swertia mussoti* dry roots using cells from *Aspergillus niger*, isolated and acylated with vinyl acetate and Novozym 435 to provide 39 (acylated only in the aromatic OH group, no yield reported), an alkaloid which anti-hepatitis and anticancer activity seems to be very promising.

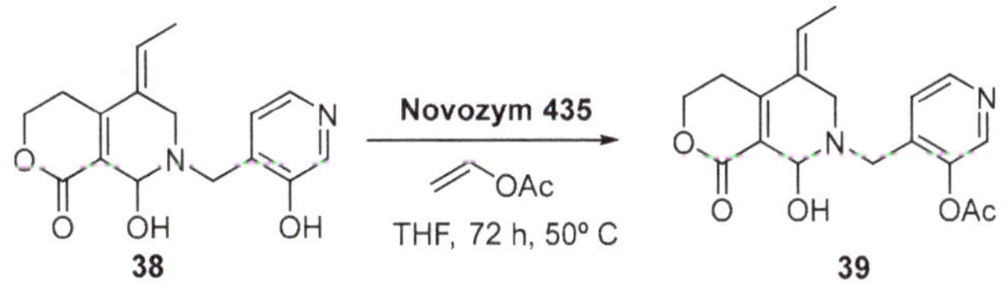

Scheme 15. Regioselective monoacylation of the iridiod-type compound **38**.

2.1.5. Other Recently-Granted Patents Using Lipases

Apart from the previous examples presented so far, some other patents have reported the use of lipases for hydrolysis [141,142] or acylations [143–147] taking advantage of the very mild reaction conditions in which they were carried out. Particularly interesting are those granted patents reporting the opening of δ-valerolactone and ε-caprolactone with mercapto-alcohols catalyzed by Novozym 435 for the preparation of polymers [148], or the lipase-catalyzed preparation of enantiopure (*R*) and (*S*)-leucine [149]. On the other hand, the in-situ reaction of vinyl acetate with isopropanol in the presence of *C. antarctica* lipase B as catalyst to obtain acetaldehyde from the vinyl alcohol (following a previously described strategy [150]) is used in a Biginelli reaction with a β-dicarbonyl compound, urea and water to produce different 3,4-dihydro-pyrimidin-2(1*H*)-ones, useful for the preparation of dark blue solid fluorescent materials [151].

To finish this section, we will illustrate an example of the well-known lipase promiscuity [68,86,152,153]. Thus, in a patent from Nanjing Tech University granted in 2018 [154], different lipases (lipase from pig pancreas (PPL) leading to the best results) were used for catalyzing the Knoevenagel condensation between indolin-2-one (Scheme 16, **40**) and different aromatic aldehydes **41** to synthesize the corresponding 3-arylidene derivatives **42**, useful in the synthesis of indole alkaloids with many therapeutic activities (antibacterial, antitumor or anti-inflammatory). Different reaction conditions were tested (solvent, temperature, etc.), being the best results obtained with a mixture of water/DMSO (1/4, v/v) and 45 °C.

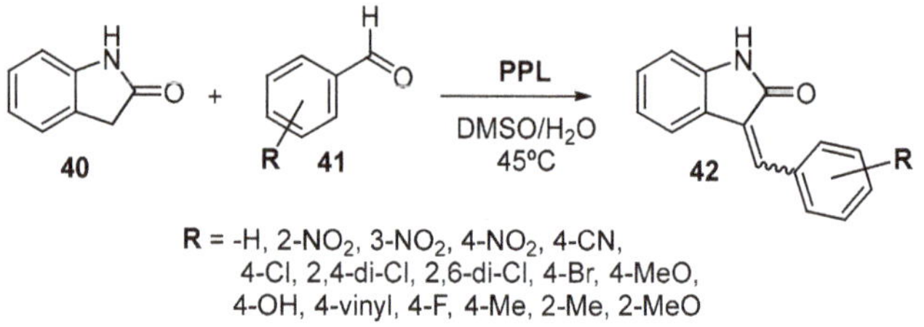
PPL
DMSO/H2O
45°C
40
41
42
R = -H, 2-NO2, 3-NO2, 4-NO2, 4-CN,
4-Cl, 2,4-di-Cl, 2,6-di-Cl, 4-Br, 4-MeO,
4-OH, 4-vinyl, 4-F, 4-Me, 2-Me, 2-MeO

Scheme 16. Lipase-catalyzed Knoevenagel condensations.

2.2. Other Hydrolases

Apart from lipases (see the previous section), other types of hydrolases have been subject of IP protection as well. For instance, novel specific sequences of esterases have been patented and successfully employed in the kinetic resolution of esters of mandelic acid [155], phenyl ethanol [156], lactate derivatives [157], or cocaine [158], as well as for the synthesis of vetiveryl esters [147]. Moreover, some esterases have been used in the penicillin production area, like in the synthesis of 3-deacetyl-7-aminocephalosporaic acid [159].

Likewise, nitrilases have been extensively characterized and protected over the last five years. Companies like BASF [160–163] and c-LEcta [164] have patented novel sequences of nitrilases, with potential use in biocatalysis in asymmetric synthesis. In addition, some other activities cover sequences applied to specific targets for pharmaceuticals and fine chemicals. For instance, nitrilases are used to generate chirality in the synthesis of precursors of L-praziquantel, a drug employed against parasitic worm infections [165,166]. Nitrilases from *Arabidopsis thaliana*, *Aspergillus niger* or *Alcaligenes faecalis* are claimed in a patent granted in 2014 from Suzhou Tongli Biomedicine Co., Ltd., to perform the kinetic resolution of nitrile **43** (Scheme 17) [167] to obtain the enantiopure (*R*)-**43**, needed for the synthesis of the required drug.

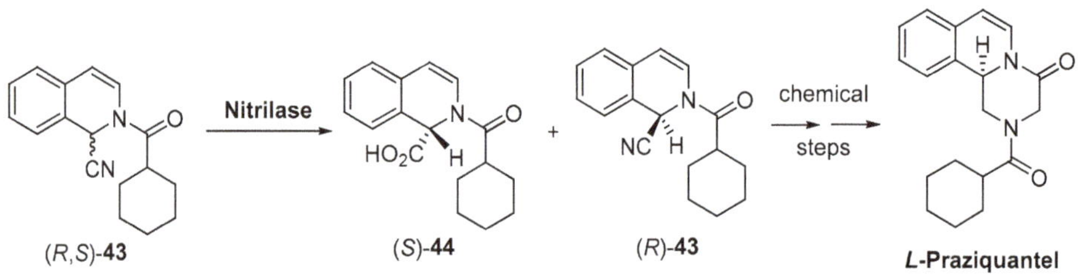
Nitrilase
chemical
steps
(R,S)-43
(S)-44
(R)-43
L-Praziquantel

Scheme 17. Use of nitrilases in the synthesis of L-praziquantel.

Using nitrilases, other patents cover the asymmetric synthesis of β-alanine from β-aminopropionitrile at high substrate loadings (> 200 g L^{-1}) [168], an improved process for the production of pregabalin (see Scheme 34, below), the active agent in Lyrica®, a drug employed in the treatment of epilepsy, neuropathic pain and fibromyalgia [169,170] starting from dinitrile substrates (developed by Hikal Ltd) [171,172], or the production of 1-cyancyclohexylacetic acid by means of engineered nitrilases, and starting from the dinitrile substrate as well [173]. In another example, the synthesis of (R)-**46**, a precursor of ivabradine (Corlanor®, Procoralan®, Coralan®, Corlentor®, Lancora® or Coraxan®, a drug prescribed for the symptomatic management of stable heart-related chest pain and heart failure [174,175]) has been developed and patented by Servier [176] (granted in 2015), using a nitrilase to generate the chirality through hydrolysis of the racemic nitrile (R,S)-**45** to furnish the carboxylic acid (R)-**46**, as depicted in Scheme 18.

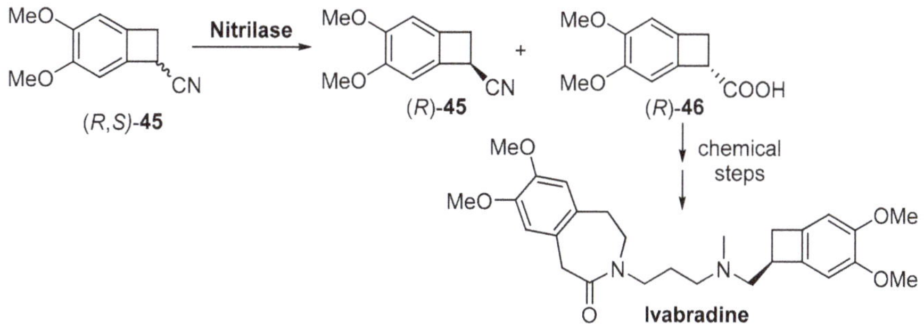

Scheme 18. Synthesis of precursors of the drug ivabradine using nitrilases.

Likewise, nitrilases have been used in the asymmetric synthesis of the statin side-chain, by hydrolysing the nitrile group [177–179]. The combination of nitrilases with halohydrin dehalogenases enables a continuous process for the efficient production of that important building block with high productivities and selectivities [179]. As reported in Section 2.1.3, the statin field is one of the most active ones when biocatalysis applied to IP generation is concerned. Other examples involving nitrilases cover the synthesis of pharmaceutical drugs like fosamprenavir [180], as well as clopidogrel [181]. Apart from nitrilases, penicillin G acylases have been used for the synthesis of optically active phenylglycine derivatives (through kinetic resolution strategies) [182].

3. Granted Patents Related to Oxidoreductases in Asymmetric Synthesis

3.1. Ketone Reductions

Alcohol dehydrogenases (ADHs; E.C. 1.1.1.x), also called ketoreductases (KREDs) or carbonyl reductases (CRs), are oxidoreductases that selectively catalyze the reversible conversion of carbonyl compounds into the corresponding alcohols, requiring the presence of nicotinamide cofactors (NADH or NADPH) to perform their activity. During the course of the reduction, the enzyme delivers a hydride from the cofactor C4 to the *Si*- (anti-Prelog ADHs) or to the *Re*-face (Prelog enzymes) of the carbonyl compound, yielding (R)- or (S)-alcohols, respectively [183–187]. Efficient cofactor regeneration methodologies are required due to the high costs of nicotinamides and inhibition processes. Thus, the biocatalyzed reduction is coupled with a secondary reaction to regenerate the nicotinamide cofactors [188–190]. In general, a second enzymatic reaction is employed, for example by combining glucose with glucose dehydrogenase (GDH), glucose-6-phosphate with glucose-6-phosphate dehydrogenase (G6PDH) or sodium formate with formate dehydrogenase (FDH). NAD(P)H can also be regenerated by using a coupled substrate approach, in which the ADH catalysed the desired bioreduction and the oxidation of a cosubstrate, in general isopropanol (IPA), which is oxidized to acetone. Ketone reduction is under thermodynamic control, so that a large excess of IPA is required.

Over the last five years, several procedures describing the use of KREDs for the preparation of valuable compounds have been patented.

Thus, a patent granted to Codexis Inc. described the preparation of ezetimibe (Zetia®, Ezetrol®, amongst other), a drug employed to treat high blood cholesterol and other lipids abnormalities [191–193], using different engineered KREDs from *Lactobacillus brevis*, *Lactobacillus kefir* and *Lactobacillus minor* in a patent granted in 2014 [194]. The ezetimibe precursor 5-((4S)-2-oxo-4-phenyl-(1,3-oxazolidin-3-yl))-1-(4-fluorophenyl)pentane-1,5-dione (**47**) was selectively reduced to (4S)-3-[(5S)-5-(4-fluorophenyl)-5-hydroxypentanoyl]-4-phenyl-1,3-oxazolidin-2-one (**48**), as depicted in Scheme 19a, with complete selectivity and conversions higher than 95% after 24 h. Glucose and glucose dehydrogenase (GHD) were employed as secondary enzymatic system to regenerate the NADP$^+$ consumed during the bioreduction. Substrate concentrations were at least of 100 g/L whereas 5 g/L of biocatalyst can be employed.

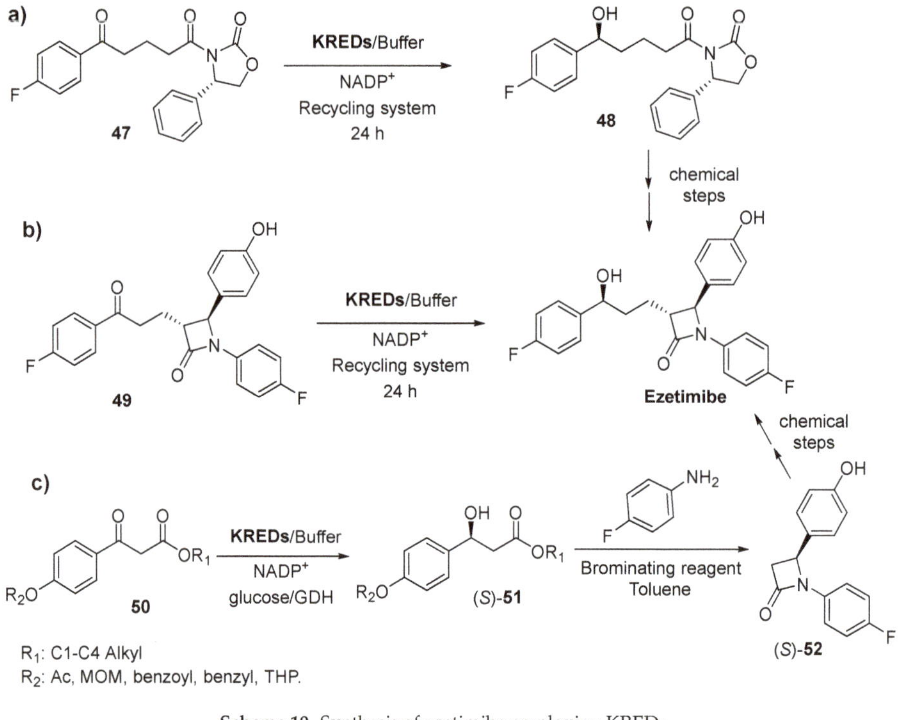

Scheme 19. Synthesis of ezetimibe employing KREDs.

The preparation of ezetimibe and some of its derivatives was also described in a further patent, also by Codexis Inc., granted in 2015 [195]. Bioreduction of 1-(4-fluorophenyl)-3(R)-[3-(4-fluorophenyl)-3-oxopropyl]-4(S)-(4-hydroxyphenyl)-2-azetidinone (**49**, Scheme 19b) was carried out by an engineered KRED from *Lactobacillus kefir*, which sequence is given. Ezetimibe was obtained with excellent diastereomeric excess (>99%) and high conversion (90%) after 24 h, being possible to use 50 g/L of substrate. Different NAD(P)$^+$ regeneration systems were employed, including a set of secondary enzymatic systems with dehydrogenases and the use of the coupled-substrate approach in presence of IPA. Ezetimibe synthesis was also described through a chemoenzymatic route in a patent granted in 2018 [196], shown in Scheme 19c. The reduction of different O-protected β-ketoesters **50** was carried out in presence of a recombinant ketoreductase leading to the β-hydroxy-esters (S)-**51** with high optical purities (>95%) and yields higher than 90% after 24 h. Both IPA as cosubstrate or glucose/GDH as secondary enzymatic system were employed for the cofactor regeneration system. The chiral β-hydroxyesters were employed to prepare an Ezetimibe intermediate (*S*)-**52** by bromination with different brominating reagents in presence of p-fluoroaniline and toluene as solvent (Scheme 19c). Immobilized KREDs have been also employed for the preparation of an ezetimibe precursor, as reported in a patent granted in 2018 [197]. Different techniques were applied to immobilize the biocatalyst, including adsorption, covalent binding, entrapping or cross-linked microencapsulated. Chiral alcohol was obtained with complete conversion after 18 h, using IPA or the glucose/GDH system for the nicotinamide cofactor regeneration.

The efficacy of statins, inhibitors of the 3-hydroxy-3-methylglutaryl-coenzyme A (HMG-CoA) reductase, against all the forms of hypercholesterolemia and for the prevention of cardiovascular events [110] has been already mentioned in Section 2.1.3. In the last years, some patents employing ketoreductases have been published for the preparation of these compounds. Thus, in 2016 a patent was granted [198] describing the use of an engineered bacterium for the preparation of (3*R*,5*S*)-6-chloro-3,5-dihydroxyhexanoic acid *tert*-butyl ester (3*R*,5*S*)-**55**, a chiral precursor of statins. The engineered bacteria contained an ADH gene from *Lactobacillus kefir* DSM20587 (ADHR), a carbonyl reductase gene and a glucose dehydrogenase gene for the nicotinamide cofactor regeneration (Scheme 20a). The bacterial resting cells were employed in the bioreduction of 6-chloro-3,5-carbonyl hexanoate (**53**) in the presence of glucose and isopropanol in order to obtain the chiral diol (3*R*,5*S*)-**55** with complete selectivity and high yields in a two-step procedure. In 2017, another patent was granted reporting the preparation of a set of statins using a chemoenzymatic methodology [199]. Thus, KRED-130, commercially available from Codexis, was able to reduce the keto group of the intermediate **56** to the corresponding chiral alcohol (*R*)-**57** with quantitative yield and optical purities higher than 99.5% (Scheme 20b). A year later, a set of diketones intermediates in the preparation of statins were selectively reduced by different KREDs to the corresponding chiral 1,3-diols in a process carried out in buffer containing some organic solvents as ethanol, acetonitrile, toluene or hexane, among others [200]. Depending on the substrate structure, the chiral diols were obtained with yields around 90% and excellent optical purities after 24 h.

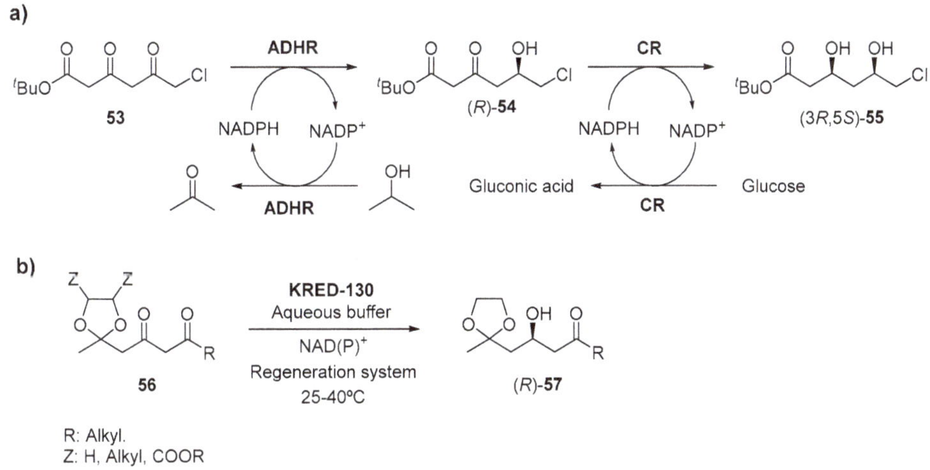

a)

b)

Scheme 20. Synthesis of statin intermediates employing ketoreductases.

Montelukast (Airon®, Everest®, Singulair®, Sansibast®, Senovital®, Accord®, amongst others) is an efficient and low toxicity anti-inflammatory, anti-allergy and asthma treatment compound [201–203]. In 2015, a patent describing the preparation of a montelukast sodium intermediate was granted [204]. The key step of this synthesis was the selective bioreduction of compound **58** (Scheme 22) catalyzed by an alcohol dehydrogenase in buffer containing an organic solvent as hexane or toluene. IPA was employed as cosubstrate in order to regenerate the NAD$^+$ cofactor. After 72 h, the chiral alcohol (*S*)-**59** can be recovered with high yield and selectivity, and converted to the desired intermediate (*S*)-**60** by treatment with methylmagnesium halide, as shown in Scheme 21. A year later, in a patent granted to Zhangjiagang Xinyi Chemical Co., a set of KREDs from Suzhou-Enzyme Biological Technology Co. were also employed in the bioreduction of the montelukast intermediate **58** to yield optically active alcohol (*S*)-**59** [205]. Bioreductions were carried out in buffer containing toluene as cosolvent using NADP+ as cofactor.

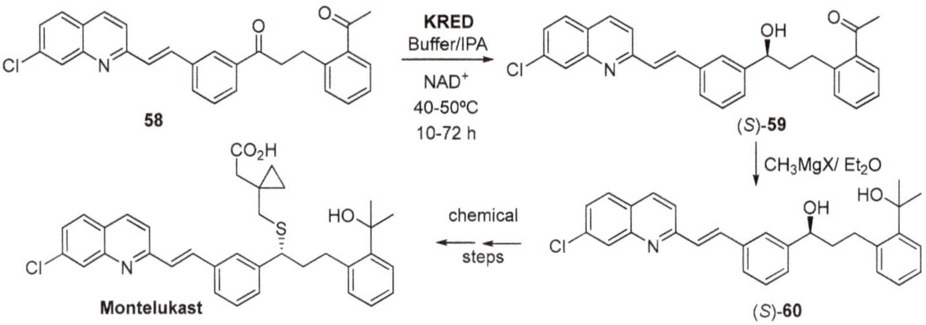

Scheme 21. Preparation of montelukast intermediates employing KREDs under mild reaction conditions.

The biocatalytic conversion of oxcarbazepine **61** and analogues to chiral (*S*)-eslicarbazepine (*S*)-**62** (a drug employed in the treatment of epilepsy [206]) and corresponding chiral alcohols is reported in a patent granted in 2015 [207] using engineered ketoreductases from *Lactobacillus*, as shown in Scheme 22. These biocatalysts were able to selectively reduce the starting ketones with optical purities and conversions higher than 90% after 24 h, depending on the substrate structure.

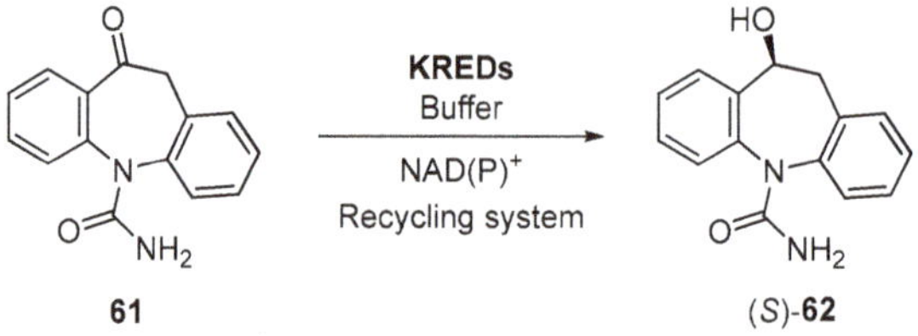

Scheme 22. Synthesis of (S)-eslicarbazepine (S)-**62** employing KREDs.

A patent covering a multistep synthesis of a set of β-3-antagonists [208] with a pyrrolidine core **65** [209] (Scheme 23) has been granted in 2017 to Merck Sharp & Dohme Corp. [210]. One of these steps comprises the use of four engineered KREDs, whose sequence is given, for the bioreduction of a set of α-aminoketones 63 to the corresponding β-aminoalcohols **64** in a dynamic kinetic resolution (Scheme 23) [211], leading to the stereoselective preparation of one of the diastereomers of the final alcohol with excellent yields and optical purities. Reactions were performed in presence of several cosolvents, whereas different enzymatic systems were tested for the nicotinamide cofactor recycling.

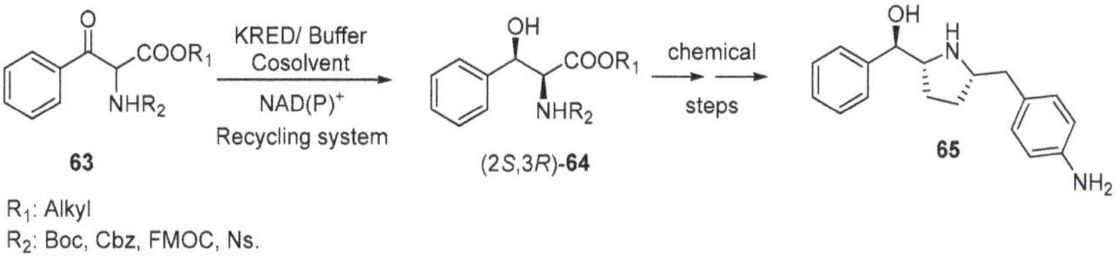

R$_1$: Alkyl
R$_2$: Boc, Cbz, FMOC, Ns.

Scheme 23. KREDs employed in the synthesis of different β-aminoalcohol intermediates of β-3-antagonists.

Dehydroepiandrosterone (DHEA, **66**) is a key intermediate in the synthesis of steroidal molecules [212]. In 2018, a granted patent included a novel approach for preparing this valuable compound by carrying out the bioreduction of the 3-oxo group of the $^5\Delta$-androstene-3,17-dione **65** in presence of KREDs obtained from *Sphingomonas wittichii* [213]. DHEA could be obtained with yields around 80–90% and selectivities higher than 90%, being only observed the reduction of the 3-oxo group, when working at substrate concentrations of 50–300 g/L (Scheme 24). Some organic solvents immiscible with water can be employed in the bioreductions in concentrations around 25–75%, and different secondary enzymatic systems were used to regenerate the NAD(P)$^+$ cofactor. The biocatalytic synthesis of DHEA has been also described in a patent granted in 2019 [214] by carrying out the same reduction at the 3-oxo positon of **65** in presence of an engineered KRED. The process was developed employing the KRED as a powder and using GDH from *Bacillus subtilis* and glucose for the NADP$^+$ recycling or using whole cells containing both enzymes. Substrate can be employed at concentrations around 120–600 g/L, furnishing enantiopure DHEA with high yields.

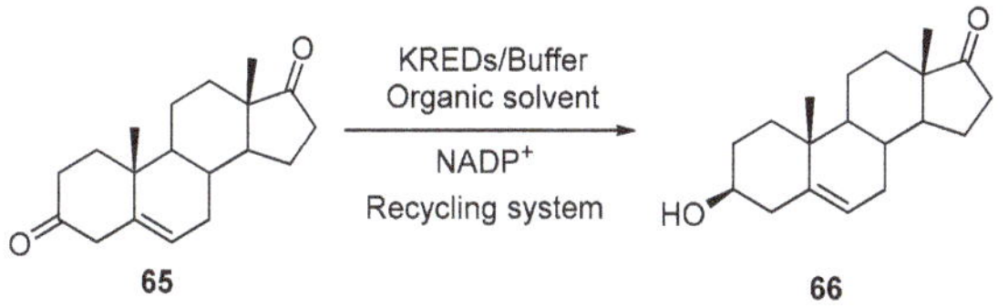

Scheme 24. Synthesis of DHEA 66 employing dehydrogenases.

A patent for the preparation of (1*R*,2*S*)-*N*-pyrrolidinyl norephedrine **68** by a bioreduction procedure has been granted to Enzymeworks in 2019 [215]. This chiral compound is a valuable synthon for the preparation of anti-AIDS drug efavirenz (Sustiva®) [216,217]. The starting ketone **67** was selectively reduced in dynamic kinetic resolution catalysed by a KRED purchased from Suzhou Chinese Biotechnology of Enzymes Co. (EW104), as indicated in Scheme 25. Isopropanol or glucose/GDH were employed for the cofactor recycling. After reaction parameters optimization, enantiopure (1*R*,2*S*)-**68** was recovered in high yields and with excellent purity.

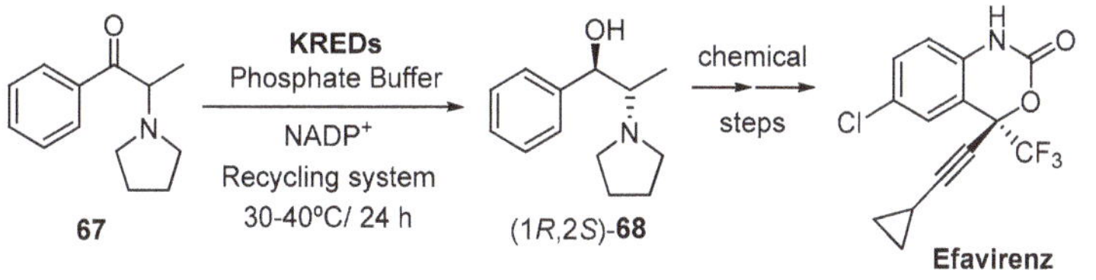

Scheme 25. Preparation of (1*R*,2*S*)-*N*-pyrrolidinyl norephedrine employing ketoreductases.

A patent granted to Goodee Pharma Co. Ltd. in 2016 described the preparation of different 3-piperidinols by employing two KREDs, which sequence is given, as biocatalysts [218]. The enzymes were employed in solution or as lyophilized powders or immobilized as free enzymes or as whole cells, being employed glucose/GDH as secondary enzymatic system for the cofactor regeneration.

The application of a set of KREDs for the enantioselective reduction of different thienyl ketones to the corresponding (*S*)-thienyl alcohols with excellent optical purities (>99% ee) and high conversions has been patented in 2018 [219]. Alcohols thus obtained are valuable chiral synthons for the preparation of duloxetine [220–222], the active enantiomer of this third generation antidepressant trademarked as Cymbalta®. A patent granted in 2015 to Codexis Inc. afforded the bioreduction of a family of 3-aryl-3-ketopropanamines to (*S*)-3-aryl-3-hydroxypropanamines employing different engineered KREDs, whose sequences are given [223]. Depending on the substrate structures, high conversions were obtained after 24 h employing glucose/GDH for the NAD(P)$^+$ recycling. The preparation of (*S*)-*N,N*-dimethyl-3-hydroxy-3-(2-thienyl)-1-propanamine ((*S*)-DHTP, **70**), a precursor in the synthesis of duloxetine, was carried out with high yields from ketone **69** (Scheme 26). (*S*)-DHTP synthesis has been also described in a patent granted to Jiangnan University in 2019 [224]. The bioreduction of the starting ketone was carried out in the presence of different ketoreductases from different microorganisms, achieving the best results using a recombinant carbonyl reductase from *Candida macedoniensis* AKU 4588. Control of the reaction conditions allows obtaining (S)-DHTP in 92% yield and 99% ee.

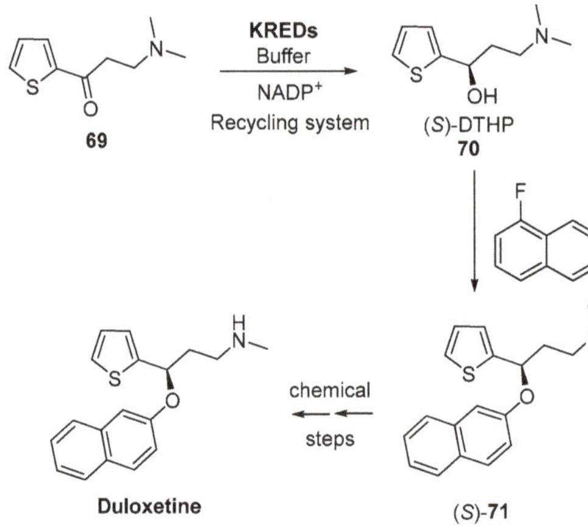

Scheme 26. Synthesis of duloxetine employing chemoenzymatic methodologies with ketoreductases.

(*R*)-2-Hydroxy-4-phenylbutyric acid ethyl ester is a valuable intermediate in the preparation of drugs employed to treat hypertension [225]. Its synthesis with a global yield of 82% starting from benzaldehyde and pyruvic acid has been patented by Suzhou Lead Biotechnology Company Ltd. [226]. The final step of this procedure comprises the bioreduction of 2-oxo-4-phenylbutyric acid ethyl ester in presence of glucose/GDH for the NADP$^+$ recycling. After 20 h at 30 °C a 97% of the enantiopure β-hydroxyester was recovered in this process. Similarly, in 2019, Genentech Inc. was granted a patent [227] for the synthesis of (*S*)-1-(1-(4-chloro-3-fluorophenyl)-2-hydroxyethyl)-4-(2-((1-methyl-1*H*-pyrazol-5-yl)amino)pyrimidin-4-yl)pyridin-2(1*H*)-one (**74**), an ERK inhibitor and a useful medicament for treating hyperproliferative disorders [228]. One of the steps for the preparation of this compound, depicted in Scheme 27, comprises the bioreduction of 1-(4-chloro-3-fluorophenyl)-2-hydroxyethanone (**72**) to (*S*)-1-(4-chloro-3-fluorophenyl)ethane-1,2-diol (*S*)-**73** catalyzed by KRED-NADH-112 (Codexis Inc.) using glucose/GDH for the NAD$^+$ recycling. The chiral alcohol was recovered with quantitative conversion and 99.5% ee.

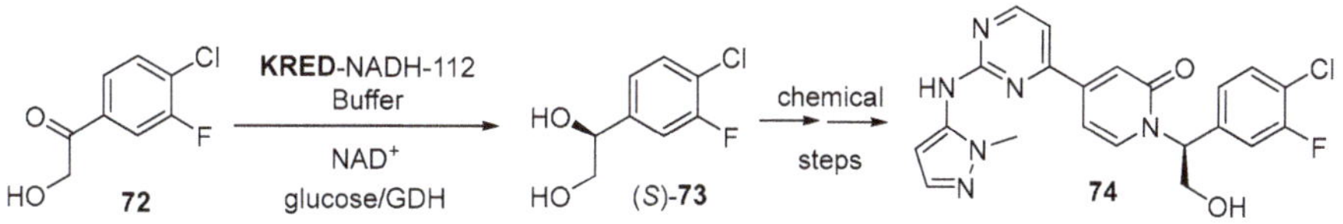

Scheme 27. The bioreduction of **72** to (*S*)-**73** in the preparation of the anticancer drug **74**.

Finally, fluoride-containing chiral alcohols are valuable compounds with a huge range of applications in medicine, pesticides or liquid crystals [229]. A patent granted to Enzymeworks in 2019 described the preparation of (*S*)-1,1,1-trifluoroisopropanol from trifluoroacetone employing different KREDs [230]. Phosphate buffer pH 7.0 was used as reaction medium containing different amounts of organic cosolvents. NADP$^+$ was recycled in presence of glucose/GDH and a fluorinated reagent was added at the end of the bioreduction in order to extract the product from the reaction media. Thus, HFE-7600 and/or F-626 were used for this purpose, being able to recover quantitatively the final alcohol with 98.2% ee. The preparation of other halogenated alcohols has been granted by Codexis Inc. in 2015 [231]. Thus, the patent described the preparation of chiral α-chloroalcohols from α-chloroketones as well as the polynucleotides encoding the engineered ketoreductases and the host cells capable of expressing the engineered ketoreductases.

3.2. Imine Reductions

In the last few years, imine reductases (IREDs) have appeared as valuable biocatalysts for the preparation of chiral amines by catalyzing the reduction of imines [232–240]. These enzymes are involved in many natural processes such as the biosynthesis of cofactors, alkaloids and cyclic amino acids. IREDs are enzymes requiring NAD(P)H, being this cofactor responsible of the imine reduction. In 2017 and 2018, two patents granted to Codexis Inc. showed different engineered enzymes with activity as imine reductases and applied these biocatalysts in the conversion of ketones and amines to the corresponding optically active secondary and tertiary amines, as for instance the conversion of cyclohexanone and L-norvaline to (*S*)-2-(cyclohexylamino)pentanoic acid [241,242]. On the other hand, a patent granted in 2016 to Pfizer Inc. [243] described the use of IRED in combination with an amine oxidase or a transaminase in one of the steps of the synthesis of pregabalin (discussed in Section 2.2, see also Scheme 34 in Section 4).

3.3. Oxidations

Oxidation processes involving different types of biocatalysts as mild oxidants under mild reaction conditions are employed are widely used nowadays while obtaining in general high selectivities [244–254]. For this reason, processes in which oxidative enzymes are employed have been patented in the last times. Most of the examples described the use of monooxygenases in different oxidations; these enzymes are able to perform the insertion of one atom of oxygen in the substrate from molecular oxygen, requiring nicotinamides as cofactors [255–259].

Esomeprazole (*S*)-**76** (Scheme 28), is a proton pump inhibitor prescribed for the treatment of dyspepsia, peptic ulcer disease and gastroesophageal reflux disease [260]. This compound is the (*S*)-enantiomer of omeprazole, which is the racemic mixture of the (*S*) and (*R*) enantiomers. Thus, in a patent granted in 2016 [261] a set of engineered cyclohexanone monooxygenases (CHMOs) from *Acinetobacter calcoaceticus* [262] were employed in the asymmetric sulfoxidation of a set of prazoles, including 5-methoxy-2-((4-methoxy-3,5-dimethylpyridin-2-yl)methylthio)-1*H*-benzodimidazole (**75**) (Scheme 28a).

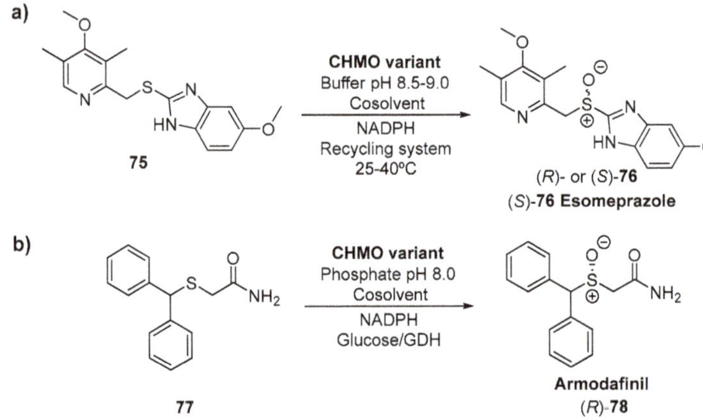

Scheme 28. Enzymatic sulfoxidations catalyzed by CHMO variants for the synthesis of esomeprazole (**a**) and armodafinil (**b**).

The oxidation of these compounds in presence of recombinant CHMOs afforded the corresponding (*R*)- or (*S*)-sulfoxides **76** with excellent optical purities at mild temperatures. In order to ensure a complete conversion, dissolved molecular oxygen in the process can be increased by bubbling the reaction with oxygen-containing gas or by use of bubble-free aeration with oxygen-containing gas. Different organic cosolvents were tested in the processes (MeOH, EtOH, IPA, acetonitrile, etc.) to increase the substrate solubility. In order to regenerate the NAD(P)H employed as cofactor, secondary enzymatic systems including glucose/GDH, glucose-6-phosphate/glucose-6-phosphate dehydrogenase (G6PDH), formate/formate dehydrogenase (FDH), phosphite/phosphite dehydrogenase (PTDH) or alcohol/ADH were used.

Engineered CHMOs have been also employed in the preparation of armodafinil ((*R*)-**78**, Scheme 28b) and analogues, as shown in a patent granted in 2017 to Codexis [263]. Armodafinil (Nuvigil®) is the active (*R*)-enantiomer of the racemic drug modafinil (Provigil®), employed for the treatment of narcolepsy and other medical conditions [264,265]. The sulfoxidation reaction was performed on both the amide substrate 2-(benzhydrylsulfinyl)acetamide **77** to obtain armodafinil, 2-(*R*)-(diphenylmethyl)sulfinylacetamide (*R*)-**78**, or on the corresponding acid substrate, benzhydryl-thioacetic acid to yield (*R*)-2-(benzhydrylsulfinyl)acetic acid, known as (*R*)-modafinil (or modafinilic acid), which can be subsequently converted to the amide product in an easy way. For both processes, high optical purities were obtained (>90% ee) under the optimized conditions when carrying out the oxidations in phosphate buffer in presence of different cosolvents.

The use of cytochrome P450 [255,256,259] has been also described in patents for the preparation of valuable compounds. Thus, in a patent granted in 2019 [266], different P450s obtained from *Bacillus subtilis* and *Bacillus cereus* expressed in *E. coli*, in combination with an electron-transfer protein having the activity to transfer an electron to the cytochrome P450, have been used as biocatalysts on the oxidation of the 3-methylene position of α-guaiene **79** to lead (-)-rotundone **80**, a molecule with several applications in the fragrance industry [267], as shown in Scheme 29.

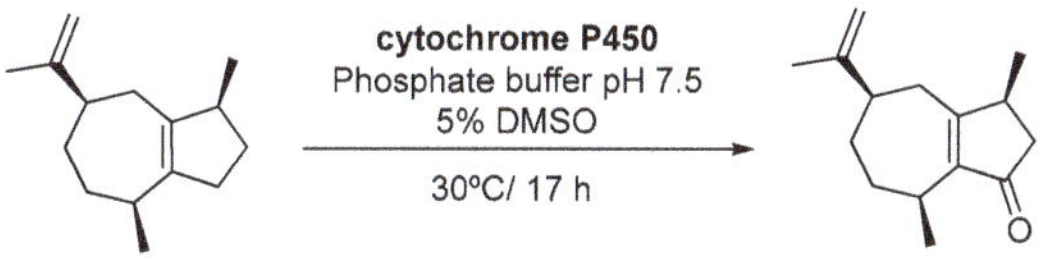

cytochrome P450
Phosphate buffer pH 7.5
5% DMSO
30°C/ 17 h
79
(-)-80

Scheme 29. Synthesis of (–)-rotundone employing *E. coli* cells expressing cytochrome P450s.

In 2015, Novozymes AS was granted a patent describing different fungal enzymes with peroxygenase activity [268]. Peroxygenases are biocatalysts capable of catalyse a very important process in organic chemistry such as the functionalisation of C–H bonds through the hydroxylation of both non-activated and aromatic C–H bonds [252,254]. The patent included the activity and compositions of these enzymes, their encoding polynucleotides, expression vectors and recombinant host cells as well as the methods of producing the enzymes. Finally, these biocatalysts were applied in the regioselective oxygenation of a set of *N*-heterocycles.

4. Granted Patents Related to Transaminases in Asymmetric Synthesis

Transaminases (TAs, type I and IV of pyridoxal 5′-phosphate (PLP)-dependent enzymes) are enzymes capable of catalysing the reversible transfer of an amino group from a suitable donor to a carbonyl acceptor. Of the two types of PLP-dependent TAs classified according to the type of substrate they convert [269], the use of α-TAs, exclusively converting α-amino and α-keto acids, is more limited, while ω-TAs can accept substrates with a distal carboxylate group. More specifically, amine TAs (ATAs), a subgroup of ω-TAs, are capable of accepting substrates not possessing a carboxylate group in their structures, have received substantial attention in recent years, because of their capability in the synthesis of chiral primary amines starting from the corresponding prochiral ketones [7,120,270].

Consequently, transaminase sequences have been protected in several granted documents. For instance, Hoffmann-La Roche and the Bornscheuer group have jointly patented (granted in 2018) several mutants of the transaminase from *Ruegeria* sp., which are useful for transamination synthetic reactions [271]. Likewise, Codexis has been active in identifying and protecting several transaminase sequences, with several patents granted between 2015–2019, with special emphasis on the production of (*R*)-ethyl-3-amino-3-(pyridine-2-yl)-propanoate derivatives [272–274]. An analogous sequence-based approach for IP construction has been successfully applied by DSM (patent granted in 2017 [275]) and Asymchem Laboratories (patent granted in 2019 [276]) with application of the novel transaminases to particularly impeded ketones as substrates (e.g., *m,m*-Cl-disubstituted phenyl-, or naphthyl acetophenones). Other examples describe the synthesis of L-aminobutyric acid, an intermediate in the synthesis of anti-epileptic levetiracetam (see also Scheme 1) in a patent granted in 2018 [277], as well as many other diverse synthetic purposes [278–280]. In the same field, the protection of novel transaminase sequences has been combined with immobilization [281–283], to be used, for instance, in the synthesis of antidiabetic sitagliptin (Januvia®, a dipeptidyl peptidase-4 (DPP-4) inhibitor which was the first marketed oral antihyperglycemic drug belonging to the gliptin family [284]) through transamination [285–287], following the pioneering example described by Merck and Codexis [288].

With respect to asymmetric synthesis, there are several remarkable examples of granted patents using transaminases for the preparation of optically active amines, very relevant building blocks for pharmaceutical chemistry as already mentioned before in Section 2.1.3 [117,118,120]. Herein, Lonza and the Bornscheuer group have jointly focused on the synthesis of *N*-amino-pyrrolidines and piperidines, using transaminases with alanine as amino donor in two patents granted between 2014 and 2017 [289,290]; thus, the formed by-product pyruvate (from alanine) can be decarboxylated by means of a pyruvate decarboxylase (PDC), or reduced by using a dehydrogenase, to shift the equilibrium towards the product formation. In another example granted in 2017 [291], transaminases have been used for the synthesis of different 3-amino-pyrrolydines **82** and **86**, starting from the correspondent functionalized oxopyrrolidines **81** and **84**, using *iso*propyl-amine as amino donor (Scheme 30). The products formed undergo subsequent cyclization, and provide access to pyrrolo[3,4-*b*]pyridine structures **83** and **86**, precursors of pharmaceuticals, e.g., the broad-spectrum antibiotic moxifloxacin (Avelox®, Vigamox® or Moxiflox®, amongst others) [292]. Remarkably, when the protecting group (PG) is an ester, the use of lipases (namely CALB) for the deprotection step is considered in the invention as well.

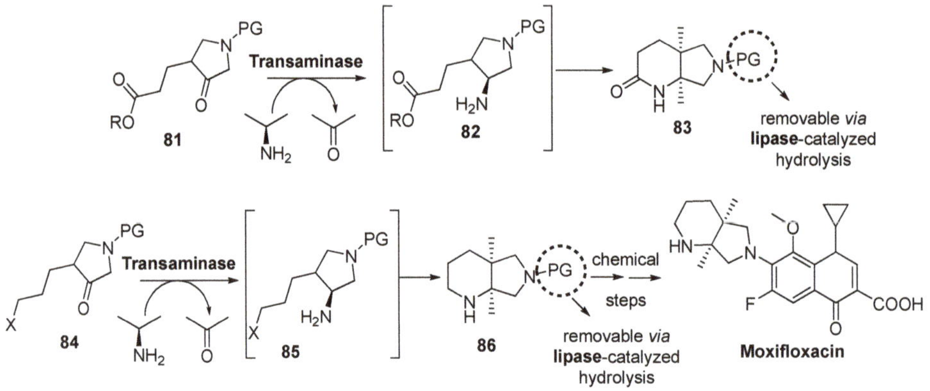

Scheme 30. Use of transaminases to afford 3-aminopyrrolidines, as precursors of pharmaceutical compounds, e.g., moxifloxacin.

Likewise, Codexis has described a patented process, granted in 2015 [293] for the synthesis of (1*R*,2*R*)-2-(3,4-dimethoxyphenethoxy)-cyclohexanamine (*R*,*R*)-**88**, starting from the correspondent enantiopure ketone (*R*)-**87** by using specific sequences of transaminases, as depicted in Scheme 31. The obtained amine is an intermediate in the synthesis of vernakalant, an ion channel blocker useful for the treatment of atrial fibrillation [294,295]. As amino donor, *iso*propyl-amine was used, and DMSO was added as co-solvent. Substrate loadings of >10 g/L have been claimed for the invention.

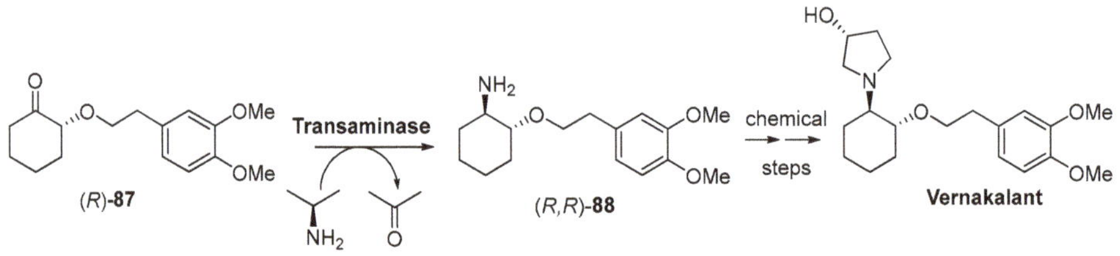

Scheme 31. Use of transaminases for pharmaceutical precursors.

Following an analogous strategy, the same company has protected other synthetic routes, such as for (*S*)-3-(1-aminoethyl)-phenol ([296], granted in 2017), the synthesis of precursors of optically active lactams ([297], granted in 2015) and the preparation of aminocyclohexyl ether compounds ([298], also granted in 2015). Other applicants have protected the synthesis of aminocyclopamine through the enzymatic transamination of the corresponding ketocyclopamines [299], or for the synthesis of spiroindolones [300]. Finally, Evonik has been active in the valorization of biogenic resources, such as isosorbide, from which other building blocks like isomannide or isoidide can be derived; in this context, the use of transaminases and dehydrogenases has been recently protected [301].

With respect to aliphatic amines, several granted patents have been found as well. Enzymeworks has covered the use of transaminases to synthetize (*S*)-2-amino-1-butanol from the corresponding ketone. Several amino donors were used, and specific transaminases protected [302]. Another patent granted in 2018 [303] protects the synthesis of (*R*,*R*)-**90**, a precursor of efinaconazole (Jublia®, Clenafin®), an anti-fungal drug [304,305], using transaminase variants able to accept the sterically hindered substrate (*R*)-**89** (Scheme 32).

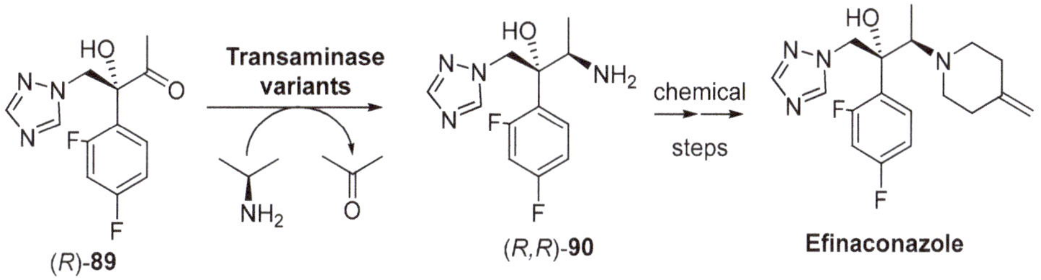

Scheme 32. Transaminase-catalyzed synthesis of efinaconazole intermediates, using transaminases.

Another relevant example is the transaminase-catalyzed synthesis of L-glufosinate, which is a broad-spectrum herbicide [306], starting from the correspondent oxoacid **91** (Scheme 33). The specific sequence of the transaminase is provided as well in the patent granted in 2018 [307].

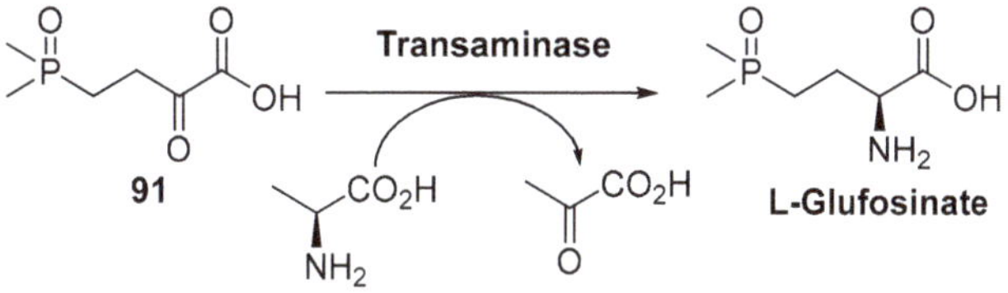

Scheme 33. Transaminase-catalyzed synthesis of L-glufosinate.

In the same field, the company Agrimetis LLC has patented a two-step enzymatic method for the synthesis of L-glufosinate, starting either from D-glufosinate or from the corresponding racemic mixture. In the first step, a D-amino acid oxidase forms the correspondent keto-acid, which is subsequently aminated by a transaminase in the second step [308]. The sequences of both enzymes are protected as well, and the immobilization of them is considered too.

Another active field is the synthesis of pregabalin (Lyrica®), a medication used for anxiety disorder, epilepsy, neurophatic pains, fibromyalgia, restless leg syndrome, etc. [169,170], already discussed in previous sections. In fact, Pfizer has patented a route ([243], granted in 2017), that covers all the synthetic steps for the generation of the amine precursor **92**, which is then subjected to a transaminase-catalyzed reaction to introduce the chirality and produce pregabalin, as depicted in Scheme 34. The use of an amine oxidase or an imine reductase is considered and protected as well.

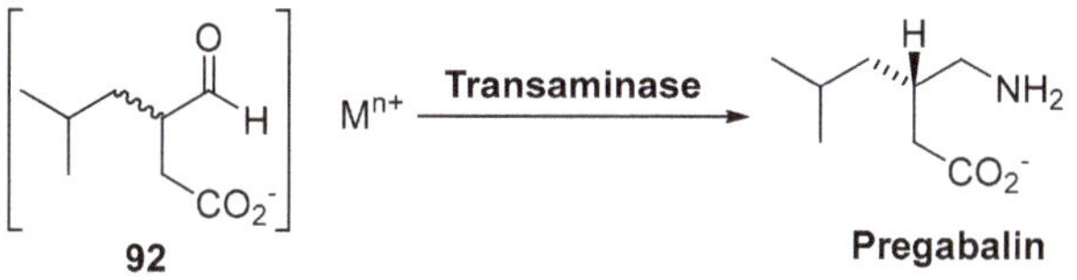

Scheme 34. Transaminase-catalyzed step for the synthesis of pregabalin, patented by Pfizer.

In an analogous area, Merck has patented a transaminase route ([309], granted in 2018) involving dynamic kinetic resolution for the synthesis of optically active amines, useful as precursors of drugs such as the anti-cancer agent niraparib [109].

As already shown in several examples (see above), the use of multi-step enzymatic processes is a rather common approach, as the properties of several biocatalysts can be combined for performing efficient asymmetric syntheses. Thus, MH2 Biochemical Ltd has patented a process for the synthesis of D-alanine, using racemases and amino transferases, being expressed as a fusion enzyme in a whole-cell [310]. In an analogous fashion, the use of lipases and transaminases are combined to deliver optically active β-amino acids, as well as precursors for the sitagliptin synthesis [311]. A further approach in this respect is the use of transaminases with opposite enantioselectivity for deracemization purposes, starting from racemic amino acids, as shown in Scheme 35 [312,313]. As depicted, the racemic (*R*,*S*)-**93** is used as amino donor by a D-amino acid transaminase, rendering the desired amino acid (*R*)-**95** from the keto acid **94**. The subsequent byproduct, the keto acid **96**, is recycled back to the racemic mixture by means of an (*S*)-transaminase, which uses *iso*propylamine as sacrificial amino donor. Overall, it represents an outstanding example on how the exquisite selectivity of enzymes may be smartly used for synthetic purposes.

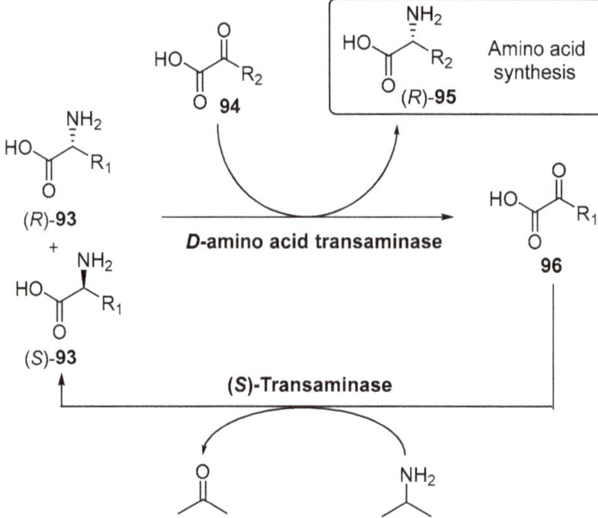

Scheme 35. Conceptual approach for production of optically active amino acids using two transaminases with opposite enantioselectivity.

5. Granted Patents Related to Lyases in Asymmetric Synthesis

Apart from other enzymes (see previous sections), lyases (EC 4.1.X.X, enzymes catalysing bond forming and breaking through non-hydrolytic or redox mechanisms) have also been matter of IP activity over the last five years, with relevant examples in the pharmaceutical and fine chemicals fields. As a common strategic working line for IP construction, several industries have focused on the (successful) protection of novel sequences of lyases (establishing several degrees of homology), which can be subsequently used for biocatalytic purposes. As relevant examples, Codexis was granted patents ([314], granted in 2019) that comprise examples of phenylammonia lyase variants (PAL, a C-N lyase [315]) with improved tolerance to pH, higher activity, or resistance to enzymatic proteolysis (key amino acidic residues of the sequence are given). In a similar approach, another patent granted in 2019 [316] protect novel another C-N lyase, tyrosine ammonia lyase variants (TAL), with analogous focus on improved activities, and including thermostability as well. Likewise, BASF (Verenium) has protected ([317], granted in 2015) the use of ammonia lyases, covering PAL and TAL variants, and broadening their invention to histidine ammonia lyases as well. With a focus on other enzyme types, in a patent granted in 2015 [318], Firmenich has covered the sequence of a 13-hydroperoxide lyase (with an alteration of more than 40 amino acids compared to the wild-type sequence). The novel variant is employed in a reaction involving several polyunsaturated fatty acids (e.g., alpha-linolenic), to yield 3-(Z)-hexen-ol. Other examples are related to novel sequences of hydroxynitrile lyases, such as a patent granted 2018 of the Asano research group [319], or another one granted in 2015 from Evocatal [320]. Similarly, in a patent granted in 2014, DSM has protected hydroxynitrile lyases that can catalyze the asymmetric synthesis of sterically hindered cyanohydrins (e.g., starting from ortho-benzaldehyde derivatives as substrates) [321].

With respect to applications in asymmetric synthesis, tyrosine phenol lyases have found applications in the field of cathecol- and phenyl- amine derivatives. For instance, Changxing Pharmaceutical Ltd company has developed in a patent granted in 2018 [322] a process for the synthesis of L-DOPA, which is used as an anti-Parkinsonian drug [323–325]. In a similar area, the Rother group has protected in two patents granted 2018 and 2017 [326,327] the synthesis of the psychoactive drug cathine ((1*S*,2*S*)-nor-pseudoephedrine) [328,329], through a multi-step enzymatic reaction combining the use of *Acetobacter pasteurianus* pyruvate decarboxylase as (*S*)-selective lyase (C-C forming type [330,331]), together with the transaminase of *Chromobacterium violaceum* as (*S*)-selective transaminase, as depicted in Scheme 36.

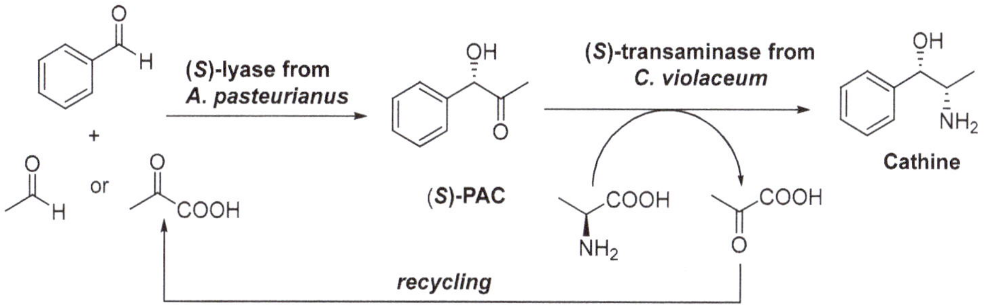

Scheme 36. Process for the synthesis of cathine, combining several enzymes and starting from inexpensive substrates.

As starting substrates, benzaldehyde and either acetaldehyde or pyruvic acid could be successfully employed. This has enabled the (re)use of pyruvic acid generated as by-product during the transaminase reaction when using L-alanine as amino donor, leading to a fully integrated multi-step enzymatic process with high selectivity and productivity to the desired compound [332,333]. The technology has been recently extended to other substrates (e.g., 2, 5-dimethoxybenzaldehyde) to afford other synthetically relevant building blocks [334].

Another relevant field of innovation with lyases (involving granted patents) is the synthesis of statins, from which many biocatalytic applications have been envisioned and comprehensively reviewed elsewhere [110], and some patents granted involving statins synthesis have been already commented on in Sections 2.1.3 and 3.1. Hence, several industries have been working on the use of lyases for this process, with different focuses to successfully build the IP. For instance, Lek Pharmaceuticals holds several patents granted in 2017 and 2018 [335–337], in which the innovation is based on the use of novel substrates, e.g., the use of 2,2-dimethoxyethanal to react with two equivalents of acetaldehyde, as well as on the use of aldolases (DERA) with novel sequences, and further enzymatic oxidation to the lactol derivative. An analogous strategy has been used by Pfizer, protecting novel DERA sequences to be used as catalysts for the reaction of *N*-protected substrates such as **99** and **100**. Specifically, the introduction of 3-phtalimidopropionaldehyde **97** and 3-succinimido-propionaldehyde **98** as substrates (Scheme 37) has been successfully protected in a patent granted in 2014 [338].

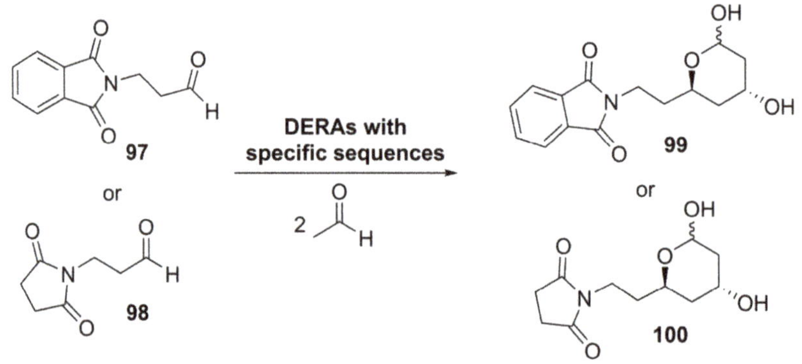

Scheme 37. Pfizer approach, using several *N*-substituted substrates, for the synthesis of statins.

In a similar approach, Mitsui Chemicals has based its innovation patented in 2016 on the protection of a DERA with a specific sequence [339], whereas QR Pharmaceuticals Ltd has patented the use of carbamates as *N*-protected substrates, to afford the synthesis of the statin side-chain using lyases ([340], granted in 2017).

Other examples of granted patents involving lyases are the synthesis of tagatose using novel variant sequences of fructose biphosphate aldolases [341,342], granted 2016 and 2015), or the synthesis of L-aminobutyric acid starting from glycine and ethanol as readily available substrates, and using a multi-sept process comprising an alcohol dehydrogenase, a threonine aldolase, a threonine deaminase, and a L-amino acid dehydrogenase ([343], granted 2018). Likewise, the Kroutil group has patented a process granted in 2016 [344,345] and presented in Scheme 38, in which *p*-vinyl- phenols **104** are produced along a three-step, one-pot reaction starting with substituted phenols **101**, pyruvic acid and ammonia as substrates. Thus, the action of a tyrosine-phenol-lyase (TPL) leads to the formation of L-tyrosine derivatives (*S*)-**102**. Remarkably, chirality generated upon furnishing the corresponding aminoacids is afterwards destroyed along the course of the process. Subsequently, phenyl-ammonia lyase (PAL or TAL) renders the substituted *p*-coumaric acids **103**, which are finally decarboxylated by a phenolic-acid decarboxylase (PAD) to afford the desired *p*-vinylphenols **104** in high yields.

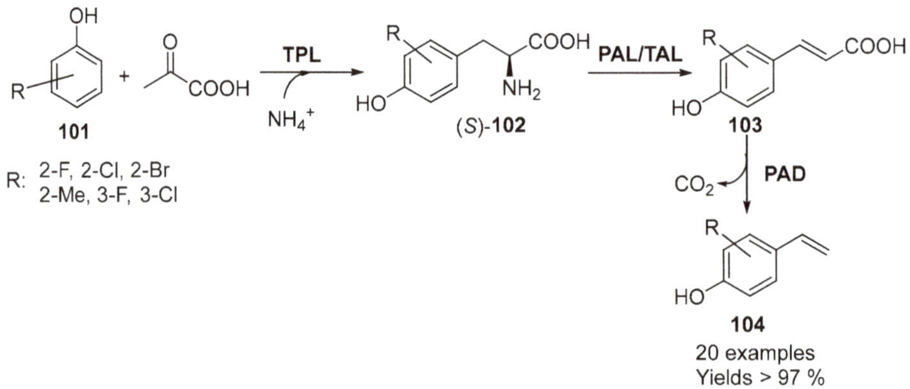

Scheme 38. Biocatalytic formation of *p*-vinylphenols **104** catalyzed by a three-step one-pot reaction starting from readily available substrates.

6. Summary and Outlook

Biocatalysis is an interdisciplinary field, in which the interaction of scientists with different backgrounds is necessary, namely, biology, chemistry, engineering, business, etc. On this basis, it offers many opportunities for innovation, and for the construction of novel IP structures. To reflect that innovation, a survey on patentability actions related to biocatalysis and asymmetric synthesis is given within this paper. To reinforce their innovative step, only granted patents have been considered. The result is a vast area of enzyme uses, covering all types, and performed by industry and academic groups. Several key ideas for innovation have been found. One is the protection of novel protein sequences for well-known reactions. Another one is the use of novel substrates for useful synthetic reactions. Further innovative lines (ending in granted patents) are the combination of different enzymes, as well as the set-up of other technical reaction parameters for improved processes. Based on the obtained results, it can be concluded that biocatalysis is taking significant steps in chemical industries, being seriously considered as a powerful alternative for combining sustainable chemistry with high efficiency and selectivity.

Author Contributions: All authors (P.D.d.M., G.d.G. and A.R.A.) equally contributed to this manuscript.

Funding: This research received no external funding.

Conflicts of Interest: The authors declare no conflict of interest.

References

1. Turner, N.J.; Kumar, R. Editorial overview: Biocatalysis and biotransformation: The golden age of biocatalysis. *Curr. Opin. Chem. Biol.* **2018**, *43*, A1–A3. [CrossRef] [PubMed]
2. Sheldon, R.A.; Brady, D. The limits to biocatalysis: Pushing the envelope. *Chem. Commun.* **2018**, *54*, 6088–6104. [CrossRef] [PubMed]
3. Rosenthal, K.; Lutz, S. Recent developments and challenges of biocatalytic processes in the pharmaceutical industry. *Curr. Opin. Green Sustain. Chem.* **2018**, *11*, 58–64. [CrossRef]
4. Raveendran, S.; Parameswaran, B.; Ummalyma, S.B.; Abraham, A.; Mathew, A.K.; Madhavan, A.; Rebello, S.; Pandey, A. Applications of microbial enzymes in food industry. *Food Technol. Biotechnol.* **2018**, *56*, 16–30. [CrossRef] [PubMed]
5. Li, G.Y.; Wang, J.B.; Reetz, M.T. Biocatalysts for the pharmaceutical industry created by structure-guided directed evolution of stereoselective enzymes. *Bioorg. Med. Chem.* **2018**, *26*, 1241–1251. [CrossRef] [PubMed]
6. Hughes, G.; Lewis, J.C. Introduction: Biocatalysis in Industry. *Chem. Rev.* **2018**, *118*, 1–3. [CrossRef] [PubMed]
7. Kelly, S.A.; Pohle, S.; Wharry, S.; Mix, S.; Allen, C.C.R.; Moody, T.S.; Gilmore, B.F. Application of Omega-Transaminases in the Pharmaceutical Industry. *Chem. Rev.* **2018**, *118*, 349–367. [CrossRef] [PubMed]
8. Hughes, D.L. Biocatalysis in drug development highlights of the recent patent literature. *Org. Process Res. Dev.* **2018**, *22*, 1063–1080. [CrossRef]
9. Dorr, B.M.; Fuerst, D.E. Enzymatic amidation for industrial applications. *Curr. Opin. Chem. Biol.* **2018**, *43*, 127–133. [CrossRef]
10. Bornscheuer, U.T. The fourth wave of biocatalysis is approaching. *Philos. Trans. R. Soc. A Math. Phys. Eng. Sci.* **2018**, *376*, 20170063. [CrossRef]
11. Chapman, J.; Ismail, A.E.; Dinu, C.Z. Industrial Applications of Enzymes: Recent Advances, Techniques, and Outlooks. *Catalysts* **2018**, *8*, 238. [CrossRef]
12. Guajardo, N.; Domínguez de María, P. Continuous biocatalysis in environmentally-friendly media: A triple synergy for future sustainable processes. *ChemCatChem* **2019**, *11*, 3128–3137. [CrossRef]
13. Woodley, J.M. Accelerating the implementation of biocatalysis in industry. *Appl. Microbiol. Biotechnol.* **2019**, *103*, 4733–4739. [CrossRef] [PubMed]
14. Sheldon, R.A.; Brady, D. Broadening the Scope of Biocatalysis in Sustainable Organic Synthesis. *ChemSusChem* **2019**, *12*, 2859–2881. [CrossRef] [PubMed]

15. Prier, C.K.; Kosjek, B. Recent preparative applications of redox enzymes. *Curr. Opin. Chem. Biol.* **2019**, *49*, 105–112. [CrossRef] [PubMed]

16. Bornscheuer, U.T.; Huisman, G.W.; Kazlauskas, R.J.; Lutz, S.; Moore, J.C.; Robins, K. Engineering the third wave of biocatalysis. *Nature* **2012**, *485*, 185–194. [CrossRef] [PubMed]

17. Foley, A.M.; Maguire, A.R. The Impact of Recent Developments in Technologies which Enable the Increased Use of Biocatalysts. *Eur. J. Org. Chem.* **2019**, *2019*, 3713–3734. [CrossRef]

18. Adams, J.P.; Brown, M.J.B.; Diaz-Rodriguez, A.; Lioyd, R.C.; Roiban, G.D. Biocatalysis: A Pharma Perspective. *Adv. Synth. Catal.* **2019**, *361*, 2421–2432. [CrossRef]

19. Domínguez de María, P.; de Gonzalo, G. *Biocatalysis: An Industrial Perspective*; Royal Society of Chemistry: London, UK, 2018.

20. Woodley, J.M. Bioprocess intensification for the effective production of chemical products. *Comput. Chem. Eng.* **2017**, *105*, 297–307. [CrossRef]

21. Ramesh, H.; Nordblad, M.; Whittall, J.; Woodley, J.M. Considerations for the Application of Process Technologies in Laboratory- and Pilot-Scale Biocatalysis for Chemical Synthesis. In *Practical Methods for Biocatalysis and Biotransformations 3*; John Wiley & Sons, Ltd.: Hoboken, NJ, USA, 2016; pp. 1–30.

22. Woodley, J.M. Scale-Up and Development of Enzyme-Based Processes for Large-Scale Synthesis Applications. In *Biocatalysis in Organic Synthesis*, 2015 ed.; Faber, K., Fessner, W.D., Turner, N.J., Eds.; Georg Thieme Verlag: Stuttgart, Germany, 2015; pp. 515–546.

23. Lindeque, R.M.; Woodley, J.M. Reactor Selection for Effective Continuous Biocatalytic Production of Pharmaceuticals. *Catalysts* **2019**, *9*, 262. [CrossRef]

24. Reetz, M.T. *Directed Evolution of Selective Enzymes: Catalysts for Organic Chemistry and Biotechnology*; Wiley-VCH: Weinheim, Germany, 2017.

25. Madhavan, A.; Sindhu, R.; Binod, P.; Sukumaran, R.K.; Pandey, A. Strategies for design of improved biocatalysts for industrial applications. *Bioresour. Technol.* **2017**, *245*, 1304–1313. [CrossRef] [PubMed]

26. Rigoldi, F.; Donini, S.; Redaelli, A.; Parisini, E.; Gautieri, A. Review: Engineering of thermostable enzymes for industrial applications. *APL Bioeng.* **2018**, *2*, 17. [CrossRef] [PubMed]

27. Zeymer, C.; Hilvert, D. Directed evolution of protein catalysts. In *Annual Review of Biochemistry*; Kornberg, R.D., Ed.; Annual Reviews: Palo Alto, CA, USA, 2018; Volume 87, pp. 131–157.

28. Arnold, F.H. Directed Evolution: Bringing New Chemistry to Life. *Angew. Chem. Int. Ed.* **2018**, *57*, 4143–4148. [CrossRef] [PubMed]

29. Bornscheuer, U.T.; Hauer, B.; Jaeger, K.E.; Schwaneberg, U. Directed evolution empowered redesign of natural proteins for the sustainable production of chemicals and pharmaceuticals. *Angew. Chem. Int. Ed.* **2019**, *58*, 36–40. [CrossRef] [PubMed]

30. Bilal, M.; Iqbal, H.M.N. Tailoring Multipurpose Biocatalysts via Protein Engineering Approaches: A Review. *Catal. Lett.* **2019**, *149*, 2204–2217. [CrossRef]

31. De Gonzalo, G.; Alcántara, A.R.; Domínguez de María, P. Cyclopentyl Methyl Ether (CPME): A versatile eco-friendly solvent for applications in biotechnology and biorefineries. *ChemSusChem* **2019**, *12*, 2083–2097. [CrossRef]

32. Shanmuganathan, S.; Natalia, D.; van den Wittenboer, A.; Kohlmann, C.; Greiner, L.; Domínguez de María, P. Enzyme-catalyzed C–C bond formation using 2-methyltetrahydrofuran (2-MTHF) as (co)solvent: efficient and bio-based alternative to DMSO and MTBE. *Green Chem.* **2010**, *12*, 2240–2245. [CrossRef]

33. Pace, V.; Hoyos, P.; Castoldi, L.; Domínguez de María, P.; Alcántara, A.R. 2-Methyltetrahydrofuran (2-MeTHF): A Biomass-Derived Solvent with Broad Application in Organic Chemistry. *ChemSusChem* **2012**, *5*, 1369–1379. [CrossRef]

34. Hernáiz, M.; Alcántara, A.R.; García, J.I.; Sinisterra, J.V. Applied Biotransformations in Green Solvents. *Chem. Eur. J.* **2010**, *16*, 9422–9437. [CrossRef]

35. Cvjetko Bubalo, M.; Vidović, S.; Radojčić Redovniković, I.; Jokić, S. Green solvents for green technologies. *J. Chem. Technol. Biotechnol.* **2015**, *90*, 1631–1639. [CrossRef]

36. Clarke, C.J.; Tu, W.-C.; Levers, O.; Bröhl, A.; Hallett, J.P. Green and Sustainable Solvents in Chemical Processes. *Chem. Rev.* **2018**, *118*, 747–800. [CrossRef] [PubMed]

37. Lomba, L.; Zuriaga, E.; Giner, B. Solvents derived from biomass and their potential as green solvents. *Curr. Opin. Green Sustain. Chem.* **2019**, *18*, 51–56. [CrossRef]

38. Sheldon, R.A. The greening of solvents: Towards sustainable organic synthesis. *Curr. Opin. Green Sustain. Chem.* **2019**, *18*, 13–19. [CrossRef]

39. Florindo, C.; Lima, F.; Ribeiro, B.D.; Marrucho, I.M. Deep eutectic solvents: Overcoming 21st century challenges. *Curr. Opin. Green Sustain. Chem.* **2019**, *18*, 31–36. [CrossRef]

40. Tomé, L.I.N.; Baião, V.; da Silva, W.; Brett, C.M.A. Deep eutectic solvents for the production and application of new materials. *Appl. Mater. Today* **2018**, *10*, 30–50. [CrossRef]

41. Guajardo, N.; Muller, C.R.; Schrebler, R.; Carlesi, C.; de Maria, P.D. Deep Eutectic Solvents for Organocatalysis, Biotransformations, and Multistep Organocatalyst/Enzyme Combinations. *ChemCatChem* **2016**, *8*, 1020–1027. [CrossRef]

42. Gotor-Fernandez, V.; Paul, C.E. Deep eutectic solvents for redox biocatalysis. *J. Biotechnol.* **2019**, *293*, 24–35. [CrossRef]

43. Paiva, A.; Matias, A.A.; Duarte, A.R.C. How do we drive deep eutectic systems towards an industrial reality? *Curr. Opin. Green Sustain. Chem.* **2018**, *11*, 81–85. [CrossRef]

44. Juneidi, I.; Hayyan, M.; Hashim, M.A. Intensification of biotransformations using deep eutectic solvents: Overview and outlook. *Process Biochem.* **2018**, *66*, 33–60. [CrossRef]

45. Maugeri, Z.; de Maria, P.D. Whole-Cell Biocatalysis in Deep-Eutectic-Solvents/Aqueous Mixtures. *ChemCatChem* **2014**, *6*, 1535–1537. [CrossRef]

46. Lin, B.; Tao, Y. Whole-cell biocatalysts by design. *Microb. Cell Factories* **2017**, *16*, 106. [CrossRef] [PubMed]

47. Wachtmeister, J.; Rother, D. Recent advances in whole cell biocatalysis techniques bridging from investigative to industrial scale. *Curr. Opin. Biotech.* **2016**, *42*, 169–177. [CrossRef] [PubMed]

48. Garzón-Posse, F.; Becerra-Figueroa, L.; Hernández-Arias, J.; Gamba-Sánchez, D. Whole cells as biocatalysts in organic transformations. *Molecules* **2018**, *23*, 1265. [CrossRef] [PubMed]

49. Grunwald, P. Immobilized Biocatalysts. *Catalysts* **2018**, *8*, 386. [CrossRef]

50. Rodrigues, R.C.; Virgen-Ortiz, J.J.; Dos Santos, J.C.S.; Berenguer-Murcia, A.; Alcantara, A.R.; Barbosa, O.; Ortiz, C.; Fernandez-Lafuente, R. Immobilization of lipases on hydrophobic supports: Immobilization mechanism, advantages, problems, and solutions. *Biotechnol. Adv.* **2019**, *37*, 746–770. [CrossRef]

51. Bernal, C.; Rodriguez, K.; Martinez, R. Integrating enzyme immobilization and protein engineering: An alternative path for the development of novel and improved industrial biocatalysts. *Biotechnol. Adv.* **2018**, *36*, 1470–1480. [CrossRef]

52. Facin, B.R.; Melchiors, M.S.; Valerio, A.; Oliveira, J.V.; de Oliveira, D. Driving immobilized lipases as biocatalysts: 10 years' state of the art and future prospects. *Ind. Eng. Chem. Res.* **2019**, *58*, 5358–5378. [CrossRef]

53. Rueda, N.; dos Santos, J.C.S.; Ortiz, C.; Torres, R.; Barbosa, O.; Rodrigues, R.C.; Berenguer-Murcia, A.; Fernandez-Lafuente, R. Chemical modification in the design of immobilized enzyme biocatalysts: Drawbacks and opportunities. *Chem. Rec.* **2016**, *16*, 1436–1455. [CrossRef]

54. DiCosimo, R.; McAuliffe, J.; Poulose, A.J.; Bohlmann, G. Industrial use of immobilized enzymes. *Chem. Soc. Rev.* **2013**, *42*, 6437–6474. [CrossRef]

55. Sheldon, R.A.; van Pelt, S. Enzyme immobilisation in biocatalysis: Why, what and how. *Chem. Soc. Rev.* **2013**, *42*, 6223–6235. [CrossRef]

56. Jesionowski, T.; Zdarta, J.; Krajewska, B. Enzyme immobilization by adsorption: A review. *Adsorption* **2014**, *20*, 801–821. [CrossRef]

57. Bilal, M.; Cui, J.D.; Iqbal, H.M.N. Tailoring enzyme microenvironment: State-of-the-art strategy to fulfill the quest for efficient bio-catalysis. *Int. J. Biol. Macromol.* **2019**, *130*, 186–196. [CrossRef] [PubMed]

58. Britton, J.; Majumdar, S.; Weiss, G.A. Continuous flow biocatalysis. *Chem. Soc. Rev.* **2018**, *47*, 5891–5918. [CrossRef] [PubMed]

59. Sheldon, R.A. The E factor 25 years on: The rise of green chemistry and sustainability. *Green Chem.* **2017**, *19*, 18–43. [CrossRef]

60. Tickner, J.A.; Becker, M. Mainstreaming green chemistry: The need for metrics. *Curr. Opin. Green Sustain. Chem.* **2016**, *1*, 1–4. [CrossRef]

61. Sheldon, R.A. Reaction efficiencies and green chemistry metrics of biotransformations. In *Biocatalysis for Green Chemistry and Chemical Process Development*; John Wiley and Sons, Inc.: Hoboken, NJ, USA, 2011; pp. 67–88.

62. De Maria, P.D.; Hollmann, F. On the (Un)greenness of Biocatalysis: Some Challenging Figures and Some Promising Options. *Front. Microbiol.* **2015**, *6*, 5.

63. Daiha, K.D.; Angeli, R.; de Oliveira, S.D.; Almeida, R.V. Are lipases still important biocatalysts? A study of scientific publications and patents for technological forecasting. *PLoS ONE* **2015**, *10*, e0131624. [CrossRef] [PubMed]

64. Hughes, D.L. Highlights of the Recent, U.S. Patent Literature: Focus on Biocatalysis. *Org. Process Res. Dev.* **2016**, *20*, 700–706. [CrossRef]

65. Bornscheuer, U.; Kazlauskas, R.J. *Hydrolases in Organic Synthesis. Regio- and Stereoselective Biotransformations,* 2nd ed.; WILEY-VCH Verlag: Weinheim, Germany, 2006.

66. Busto, E.; Gotor-Fernandez, V.; Gotor, V. Hydrolases: Catalytically promiscuous enzymes for non-conventional reactions in organic synthesis. *Chem. Soc. Rev.* **2010**, *39*, 4504–4523. [CrossRef]

67. Siirola, E.; Frank, A.; Grogan, G.; Kroutil, W. C-C hydrolases for biocatalysis. *Adv. Synth. Catal.* **2013**, *355*, 1677–1691. [CrossRef]

68. Lopez-Iglesias, M.; Gotor-Fernandez, V. Recent advances in biocatalytic promiscuity: Hydrolase-catalyzed reactions for nonconventional transformations. *Chem. Rec.* **2015**, *15*, 743–759. [CrossRef]

69. Mendez-Sanchez, D.; Lopez-Iglesias, M.; Gotor-Fernandez, V. Hydrolases in organic chemistry. Recent achievements in the synthesis of pharmaceuticals. *Curr. Org. Chem.* **2016**, *20*, 1186–1203. [CrossRef]

70. Kazlauskas, R. Chapter 5-Hydrolysis and formation of carboxylic acid and alcohol derivatives. In *Organic Synthesis Using Biocatalysis*; Goswami, A., Stewart, J.D., Eds.; Academic Press: Cambridge, MA, USA, 2016; pp. 127–148.

71. Bertau, M.; Jeromin, G.E. Hydrolysis and transacylation: Esterases, lipases, phosphatases, and phosphoryl transferases. In *Biocatalysis in Organic Synthesis*, 2015 ed.; Faber, K., Fessner, W.D., Turner, N.J., Eds.; Georg Thieme Verlag: Stuttgart, Germany, 2015; Volume 1, pp. 129–187.

72. Liu, X.; Kokare, C. Microbial enzymes of use in Industry. In *Biotechnology of Microbial Enzymes*; Brahmachari, G., Ed.; Academic Press, Elsevier: Amsterdam, The Netherlands, 2017; pp. 267–298.

73. Sanchez, S.; Demain, A.L. Useful Microbial Enzymes—An Introduction. In *Biotechnology of Microbial Enzymes*; Brahmachari, G., Ed.; Academic Press, Elsevier: Amsterdam, The Netherlands, 2017; pp. 1–11.

74. Reetz, M.T. Lipases as practical biocatalysts. *Curr. Opin. Chem. Biol.* **2002**, *6*, 145–150. [CrossRef]

75. Roy, I.; Mukherjee, J.; Gupta, M.N. High Activity Preparations of Lipases and Proteases for Catalysis in Low Water Containing Organic Solvents and Ionic Liquids. In *Immobilization of Enzymes and Cells*, 3rd ed.; Guisan, J.M., Ed.; Humana Press: Totowa, NJ, USA, 2013; Volume 1051, pp. 275–284.

76. De Miranda, A.S.; Miranda, L.S.M.; de Souza, R. Lipases: Valuable catalysts for dynamic kinetic resolutions. *Biotechnol. Adv.* **2015**, *33*, 372–393. [CrossRef] [PubMed]

77. Seddigi, Z.S.; Malik, M.S.; Ahmed, S.A.; Babalghith, A.O.; Kamal, A. Lipases in asymmetric transformations: Recent advances in classical kinetic resolution and lipase–metal combinations for dynamic processes. *Co-ord. Chem. Rev.* **2017**, *348*, 54–70. [CrossRef]

78. Hills, G. Industrial use of lipases to produce fatty acid esters. *Eur. J. Lipid Sci. Technol.* **2003**, *105*, 601–607. [CrossRef]

79. Bornscheuer, U.T.; Flickinger, M.C. Lipases, synthesis of chiral compounds, aqueous and organic solvents. In *Encyclopedia of Industrial Biotechnology*; John Wiley & Sons, Inc.: Hoboken, NJ, USA, 2009; Volume 5, pp. 1–12.

80. Tufvesson, P.; Tornvall, U.; Carvalho, J.; Karlsson, A.J.; Hatti-Kaul, R. Towards a cost-effective immobilized lipase for specialty chemicals. *J. Mol. Catal. B Enzym.* **2011**, *68*, 200–205. [CrossRef]

81. Ansorge-Schumacher, M.B.; Thum, O. Immobilised lipases in the cosmetics industry. *Chem. Soc. Rev.* **2013**, *42*, 6475–6490. [CrossRef]

82. Adlercreutz, P. Immobilisation and application of lipases in organic media. *Chem. Soc. Rev.* **2013**, *42*, 6406–6436. [CrossRef]

83. Kumar, A.; Dhar, K.; Kanwar, S.S.; Arora, P.K. Lipase catalysis in organic solvents: Advantages and applications. *Biol. Proced. Online* **2016**, *18*, 2. [CrossRef]

84. Priyanka, P.; Tan, Y.Q.; Kinsella, G.K.; Henehan, G.T.; Ryan, B.J. Solvent stable microbial lipases: Current understanding and biotechnological applications. *Biotechnol. Lett.* **2019**, *41*, 203–220. [CrossRef] [PubMed]

85. Chakravorty, D.; Parameswaran, S.; Dubey, V.K.; Patra, S. Unraveling the rationale behind organic solvent stability of lipases. *Appl. Biochem. Biotechnol.* **2012**, *167*, 439–461. [CrossRef] [PubMed]

86. Dwivedee, B.P.; Soni, S.; Sharma, M.; Bhaumik, J.; Laha, J.K.; Banerjee, U.C. Promiscuity of lipase-catalyzed reactions for organic synthesis: A recent update. *ChemistrySelect* **2018**, *3*, 2441–2466. [CrossRef]

87. Ji, Y.; Chen, B.; Qian, F.; He, Y.; Gao, X.; Hong, X. A Method for Preparing Levetiracetam. CN105063120B, 18 October 2015.

88. Ortiz, C.; Ferreira, M.L.; Barbosa, O.; dos Santos, J.C.S.; Rodrigues, R.C.; Berenguer-Murcia, A.; Briand, L.E.; Fernandez-Lafuente, R. Novozym 435: The perfect lipase immobilized biocatalyst? *Catal. Sci. Technol.* **2019**, *9*, 2380–2420. [CrossRef]

89. Ueda, T.; Abe, Y. Novel Method for Producing 1-(acyloxy)alkyl Carbamate Derivatives. WO2016208709A1, 29 December 2016.

90. Boffa, M.B.; Koschinsky, M.L. Curiouser and curiouser: Recent advances in measurement of thrombin-activatable fibirinolysis inhibitor (TAFI) and in understanding its molecular genetics, gene regulation, and biological roles. *Clin. Biochem.* **2007**, *40*, 431–442. [CrossRef] [PubMed]

91. Leurs, J.; Hendriks, D. Carboxypeptidase U (TAFla): A metallocarboxypeptidase with a distinct role in haemostasis and a possible risk factor for thrombotic disease. *Thromb. Haemost.* **2005**, *94*, 471–487. [CrossRef] [PubMed]

92. Colucci, M.; Semeraro, N. Thrombin activatable fibrinolysis inhibitor: At the nexus of fibrinolysis and inflammation. *Thromb. Res.* **2012**, *129*, 314–319. [CrossRef]

93. Zheng, G.; Chen, Q. Method for producing ticagrelor chiral drug intermediates by using Candida antarctica lipase B. CN104164469A, 26 November 2014.

94. Husted, S.; van Giezen, J.J.J. Ticagrelor: The first reversibly binding oral P2Y (12) receptor antagonist. *Cardiovasc. Ther.* **2009**, *27*, 259–274. [CrossRef]

95. Danielak, D.; Karazniewicz-Lada, M.; Glowka, F. Ticagrelor in modern cardiology-an up-to-date review of most important aspects of ticagrelor pharmacotherapy. *Expert Opin. Pharmacother.* **2018**, *19*, 103–112. [CrossRef]

96. Abrahamson, M.J.; Kielbus, A.B.; Riordan, W.T.; Hill, D.R.; Chemburkar, S.R.; Reddy, R.E.; Towne, T.B.; Mei, J.; Brown, G.J.; Mix, S. Enzymatic process for the preparation of (1S,2R)-2-(Difluoromethyl)-1-(propoxycarbonyl) cyclopropanecarboxylic Acid. US10316338B1, 11 June 2019.

97. Markham, A.; Goa, K.L. Valsartan-A review of its pharmacology and therapeutic use in essential hypertension. *Drugs* **1997**, *54*, 299–311. [CrossRef]

98. Xia, J.; Chen, D. Method for Preparing L-Valsartan by Separating DL-Valsartan Ester with Lipase. CN105420338A, 23 March 2016.

99. Xin, J.; Sun, L.; Wang, Y.; Chen, L. Preparation Method of 1,4-Dihydro-2,6-dimethyl-4-(3-nitrophenyl)-3, 5-pyridinedicarboxylic Acid Monomethyl Ester. CN106279000A, 4 January 2017.

100. Curtis, T.M.; Scholfield, C.N. Nifedipine blocks Ca^{2+} store refilling through a pathway not involving L-type Ca2+ channels in rabbit arterioral smooth muscle. *J. Physiol.* **2001**, *532*, 609–623. [CrossRef] [PubMed]

101. Hughes, A. Calcium Channel Blockers. In *Hypertension: A Companion to Braunwald's Heart Disease*, 3rd ed.; Bakris, G.L., Sorrentino, M.J., Eds.; Elsevier: Amsterdam, The Netherlands, 2018; pp. 242–253.

102. Chen, X.; Sun, D.; Sun, Q.; He, S.; Wang, C.; Zhuang, M.; Yang, Y.; Xie, K.; Guo, P. Preparation Method of High-Purity Ledipasvir Intermediate (1R,3S,4S)-2-(tert-Butoxycarbonyl)-2-azabicyclo[2.2.1]heptane-3 -carboxylic Acid. CN105461606A, 6 April 2016.

103. Link, J.O.; Taylor, J.G.; Xu, L.H.; Mitchell, M.; Guo, H.Y.; Liu, H.T.; Kato, D.; Kirschberg, T.; Sun, J.Y.; Squires, N.; et al. Discovery of Ledipasvir (GS-5885): A potent, once-daily oral NS5A Inhibitor for the treatment of hepatitis C virus infection. *J. Med. Chem.* **2014**, *57*, 2033–2046. [CrossRef] [PubMed]

104. Ku, L. Process for Synthesis of Posaconazole Intermediate 2-Methylpropionic Acid 4-(2,4-Difluorophenyl)-2-(2S)-hydroxymethyl-4-pentenyl Ester. CN105753693A, 13 July 2016.

105. Saksena, A.K.; Girijavallabhan, V.M.; Lovey, R.G.; Pike, R.E.; Wang, H.Y.; Ganguly, A.K.; Morgan, B.; Zaks, A.; Puar, M.S. Highly stereoselective access to novel 2,2,4-trisubstituted tetrahydrofurans by halocyclization-Practical chemoenzymatic synthesis of SCH-51048, a broad-spectrum orally-active antifungal agent. *Tetrahedron Lett.* **1995**, *36*, 1787–1790. [CrossRef]

106. Oberhuber, M.; Salchenegger, J.; De Souza, D.; Albert, M.; Wilhelm, T.; Langner, M.; Sturm, H.; Spitzenstaetter, H.P. Process for the Preparation of Chiral Triazolones. WO2011144653A1, 24 November 2011.

107. Langner, M.; De Souza, D.; Pise, A.C.; Bhuta, S. Method for Preparation and Purification of Posaconazole and Posaconazole Intermediates. WO2011144657A1, 24 November 2011.

108. Kunic Tesovic, B. Purification of Posaconazole Intermediates. EP2789610A1, 2014.

109. Hoyos, P.; Pace, V.; Alcántara, A.R. Chiral building blocks for drugs synthesis via biotransformations. In *Asymmetric Synthesis of Drugs and Natural Products*; Nag, A., Ed.; CRC Press: Boca Raton, FL, USA, 2018; pp. 346–448.

110. Hoyos, P.; Pace, V.; Alcántara, A.R. Biocatalyzed synthesis of statins: A sustainable strategy for the preparation of valuable drugs. *Catalysts* **2019**, *9*, 260. [CrossRef]

111. Yu, C.; Zhai, M.; Wang, M.; Gong, X.; Zhang, T.; Zhai, H. Preparation Method of High-Purity Pitavastatin Calcium for Treating Hypercholesterolemia. CN103834705A, 15 October 2014.

112. Torres, F.; Rubin, L.J. Treprostinil for the treatment of pulmonary arterial hypertension. *Expert Rev. Cardiovasc. Ther.* **2013**, *11*, 13–25. [CrossRef] [PubMed]

113. Moriarty, R.M.; Rani, N.; Enache, L.A.; Rao, M.S.; Batra, H.; Guo, L.; Penmasta, R.A.; Staszewski, J.P.; Tuladhar, S.M.; Prakash, O.; et al. The intramolecular asymmetric Pauson-Khand cyclization as a novel and general stereoselective route to benzindene prostacyclins: Synthesis of UT-15 (treprostinil). *J. Org. Chem.* **2004**, *69*, 1890–1902. [CrossRef]

114. Wang, Q.; Wang, T.; Sun, Y.; Wang, B.; You, B.; Pu, J.; Li, X.; Jiang, Y. Preparation of Levocarnitine. CN106748843A, 31 May 2017.

115. Kraemer, W.J.; Volek, J.S.; Dunn-Lewis, C. L-Carnitine supplementation: Influence upon physiological function. *Curr. Sports Med. Rep.* **2008**, *7*, 218–223. [CrossRef]

116. Bloomer, R.J.; Butawan, M.; Farney, T.M.; McAllister, M.J. *An Overview of the Dietary Ingredient Carnitine*; Academic Press Ltd.: London, UK, 2019; pp. 605–617.

117. Gröger, H.J.A.M. Biocatalytic concepts for synthesizing amine bulk chemicals: Recent approaches towards linear and cyclic aliphatic primary amines and ω-substituted derivatives thereof. *Appl. Microbiol. Biotechnol.* **2019**, *103*, 83–95. [CrossRef]

118. Kohls, H.; Steffen-Munsberg, F.; Höhne, M. Recent achievements in developing the biocatalytic toolbox for chiral amine synthesis. *Curr. Opin. Chem. Biol.* **2014**, *19*, 180–192. [CrossRef]

119. Grogan, G. Synthesis of chiral amines using redox biocatalysis. *Curr. Opin. Chem. Biol.* **2018**, *43*, 15–22. [CrossRef] [PubMed]

120. Gomm, A.; O'Reilly, E. Transaminases for chiral amine synthesis. *Curr. Opin. Chem. Biol.* **2018**, *43*, 106–112. [CrossRef] [PubMed]

121. Ismail, H.; Lau, R.M.; van Rantwijk, F.; Sheldon, R.A. Fully enzymatic resolution of chiral amines: Acylation and deacylation in the presence of Candida antarctica lipase B. *Adv. Synth. Catal.* **2008**, *350*, 1511–1516. [CrossRef]

122. Bhardwaj, K.K.; Gupta, R. Synthesis of chirally pure enantiomers by lipase. *J. Oleo Sci.* **2017**, *66*, 1073–1084. [CrossRef] [PubMed]

123. Wang, J. Method for Preparation of (R)-2-Tetrahydronaphthylamine. CN104263801A, 7 January 2015.

124. Bruinvels, J. Evidence for inhibition of reuptake of 5-hydroxytryptamine and noradrenaline by tetrahydronaphthylamine in rat brain. *Br. J. Pharmacol.* **1971**, *42*, 281–286. [CrossRef]

125. Chen, Y. A Preparation Method of R-1-Aminoindane with Candida Rugosa Lipase as Biol. Resoln. Catalyzer. CN105063161A, 18 November 2015.

126. Oldfield, V.; Keating, G.M.; Perry, C.M. Rasagiline-A review of its use in the management of Parkinson's disease. *Drugs* **2007**, *67*, 1725–1747. [CrossRef] [PubMed]

127. McCormack, P.L. Rasagiline: A Review of Its Use in the Treatment of Idiopathic Parkinson's Disease. *CNS Drugs* **2014**, *28*, 1083–1097. [CrossRef]

128. Chen, Y. A Resolution Method for Preparing Optically Pure R-1-phenylethylamine. CN104152525A, 19 November 2014.

129. Gaboardi, M.; Pallanza, G.; Baratella, M.; Castaldi, G.; Castaldi, M. Chemoenzymic Preparation of Sofosbuvir Involving a Lipase Catalyzed Regioselective Deacetylation. WO2017144423A1, 31 August 2017.

130. Lawitz, E.; Mangia, A.; Wyles, D.; Rodriguez-Torres, M.; Hassanein, T.; Gordon, S.C.; Schultz, M.; Davis, M.N.; Kayali, Z.; Reddy, K.R.; et al. Sofosbuvir for previously untreated chronic hepatitis C infection. *N. Engl. J. Med.* **2013**, *368*, 1878–1887. [CrossRef]

131. Du, Q.H.; Peng, C.; Zhang, H. Polydatin: A review of pharmacology and pharmacokinetics. *Pharm. Biol.* **2013**, *51*, 1347–1354. [CrossRef]

132. Sohretoglu, D.; Baran, M.Y.; Arroo, R.; Kuruuzum-Uz, A. Recent advances in chemistry, therapeutic properties and sources of polydatin. *Phytochem. Rev.* **2018**, *17*, 973–1005. [CrossRef]

133. Wang, C.; Du, W.; Bi, Y.; Yuan, X.; Yang, R.; Ding, C.; Zhao, X.; Zhou, W. Preparation of Polydatin Ester Derivative and Application Thereof. CN105503970A, 20 April 2016.

134. Alcántara, A.R.; Domínguez de Maria, P. Recent advances on the use of 2-methyltetrahydrofuran (2-MeTHF) in biotransformations. *Curr. Green Chem.* **2018**, *5*, 85–102. [CrossRef]

135. Zhang, A.; Sun, H.; Sun, W.; Wang, X. Metabolomics and proteomics annotate therapeutic mechanisms of geniposide. In *Chinmedomics: The Integration of Serum Pharmacochemistry and Metabolomics to Elucidate the Scientific Value of Traditional Chinese Medicine*; Elsevier Inc.: Amsterdam, The Netherlands, 2015; pp. 157–173.

136. Yao, Z.; Lu, Y.; Ni, F.; Zhu, B.; Sun, Y. Method for Catalytic Synthesis of C-6'-Lauroyl Geniposide by Lipase. CN106282272A, 4 January 2017.

137. Leong, X.Y.; Thanikachalam, P.V.; Pandey, M.; Ramamurthy, S. A systematic review of the protective role of swertiamarin in cardiac and metabolic diseases. *Biomed. Pharmacother.* **2016**, *84*, 1051–1060. [CrossRef] [PubMed]

138. Patel, N.; Tyagi, R.K.; Tandel, N.; Garg, N.K.; Soni, N. The molecular targets of swertiamarin and its derivatives confer anti-diabetic and anti-hyperlipidemic effects. *Curr. Drug Targets* **2018**, *19*, 1958–1967. [CrossRef] [PubMed]

139. Jun, C.; Xue-Ming, Z.; Chang-Xiao, L.; Tie-Jun, Z. Structure elucidation of metabolites of swertiamarin produced by Aspergillus niger. *J. Mol. Struct.* **2008**, *878*, 22–25. [CrossRef]

140. Chang, J.; Li, Y.; Guo, J.; Yao, L. Preparation of 3-(3-acetyl-4-methylpyridine)-NHHP. CN106543193A, 29 March 2017.

141. Paio, A.; Fogal, S.; Motterle, R. Enzymatic process for the preparation of testosterone and esters thereof. EP3064589A1, 2016.

142. Wang, Z.; Yan, J.; Hong, H.; Lin, Y. Process for Preparation of (3S)-5-amino-3-[[(1,1-dimethylethyl) dimethylsilyl]oxy]-5-oxo-pentanoic Acid. CN104356155A, 18 February 2015.

143. Stensrud, K.; Venkitasubramanian, P. Esterification of 2,5-furan-dicarboxylic acid. US20150315166A1, 5 November 2015.

144. Yamashita, M.; Kinsho, T. Chemoenzymic Preparation of (2R,12Z)-2-Benzoyloxy-12-heptadecene and (2S,12Z)-2-Hydroxy-12-heptadecene. US20160076063A1, 17 March 2016.

145. Yuan, Z.; Wang, Z.; Lv, P.; Luo, W.; He, D.; Li, Z.; Fu, J.; Li, H.; Miao, C.; Yang, L. A Method for Preparing Kojic Acid Dipalmitate by Composite Enzymatic Method. CN105296554A, 3 February 2016.

146. Li, M.; Li, Z. Enzymatic Resolution of Dl-Menthol with Ionic Liquid as Green Medium. CN104531823A, 22 April 2015.

147. Antoniotti, S.; Filippi, J.J.; Notar Francesco, I.; Ramilijaona, J. Biotechnological Manufacture of Vetiveryl Esters. WO2016193208A1, 8 December 2016.

148. Guo, K.; Huang, W.; Zhu, N.; Hu, X.; Fang, Z.; Liu, Y. Method for Using Microreactor to Prepare Mercapto Functionalized Polylactone. CN105969816A, 28 September 2016.

149. Jiang, L.; Fu, Q.; He, H.; Tang, S.; Gu, M. Enzymic Resolution of Racemic Leucine. CN103981248A, 13 August 2014.

150. Perez-Sanchez, M.; Dominguez de Maria, P. Lipase Catalyzed in Situ Production of Acetaldehyde: A Controllable and Mild Strategy for Multi-Step Reactions. *ChemCatChem* **2012**, *4*, 617–619. [CrossRef]

151. Yu, X.; Zhang, W.; Wang, N. Preparation of 3,4-Dihydro-pyrimidin-2(1H)-one Derivative and Application thereof. CN106588782A, 26 April 2017.

152. Kapoor, M.; Gupta, M.N. Lipase promiscuity and its biochemical applications. *Process. Biochem.* **2012**, *47*, 555–569. [CrossRef]

153. Arora, B.; Mukherjee, J.; Gupta, M. Enzyme promiscuity: Using the dark side of enzyme specificity in white biotechnology. *Sustain. Chem. Process.* **2014**, *2*, 25. [CrossRef]

154. Hu, Y.; Huang, H.; Ding, Y.; Gu, M.; Li, H.; Ni, X. Preparation of 3-Substituted-2-Indolinone Compound with Lipase. CN104818305A, 5 August 2015.

155. Hu, Y.; Liang, J.; Sun, A.; Zhang, Y.; Deng, D. A Method for Resolution of Methyl (±)-Mandelate with Esterase. CN104830944A, 12 August 2015.

156. Hu, Y.; Liang, J.; Zhang, Y.; Sun, A.; Deng, D. Bacillus Esterase BSE01281 and Its Application in Resolution of (±)-1-Phenylethanol and (±)-Styralyl Acetate. CN104962533A, 7 October 2015.

157. Hu, Y.; Wang, Y.; Zhang, Y. Esterase PHE1414 and Its Coding Gene, and Application as Catalyst in the Preparation of Thereof. CN105802935A, 27 July 2016.

158. Zhan, C.-G.; Zheng, F.; Fang, L. High Activity Variants of Cocaine Esterase for Cocaine Hydrolysis in the Treatment of Overdose. US20160122732A1, 5 May 2016.

159. Hou, H.; Wei, W.; Wang, M.; Cheng, J.; Wang, G.; Hu, B.; Li, J.; Yu, P. Enzymatic Preparation Method of 3-Deacetyl-7-Aminocephalosporanic Acid (D-7-Aca) from Cephalosporin C (Cpc) Sodium Salt by Utilizing Immobilized Cpc Acylase and Deacetylase. CN104480181A, 1 April 2015.

160. Wong, K.; Short, J.M.; Burk, M.J.; Desantis, G.; Farwell, R. Nitrilases, Nucleic Acids Encoding Them and Methods for Making and Using Them. US9315792B2, 19 April 2016.

161. Burk, M.; Chaplin, J.A.; Chi, E.; DeSantis, G.; Milan, A.; Robertson, D.; Short, J.M.; Weiner, D.P. Nitrilases. AU2013200739B2, 7 March 2013.

162. Chaplin, J.A.; Weiner, D.P.; Milan, A.; Chi, E.; Short, J.M.; Madden, M.; Burk, M.; Robertson, D.; DeSantis, G. Nitrilases. EP2327767A1, 1 June 2011.

163. Weiner, D.P.; Chi, E.; Chaplin, J.A.; Milan, A.; Short, J.M.; DeSantis, G.; Madden, M.; Burk, M.; Robertson, D.E. Nitrilases. US8906663B2, 9 December 2014.

164. Vogel, A.; Schwarze, D.; Greiner-Stoeffele, T. Nitrilases. US8916364B2, 23 December 2014.

165. Doenhoff, M.J.; Cioli, D.; Utzinger, J. Praziquantel: Mechanisms of action, resistance and new derivatives for schistosomiasis. *Curr. Opin. Infect. Dis.* **2008**, *21*, 659–667. [CrossRef]

166. da Silva, V.B.R.; Boucherle, B.; El-Methni, J.; Hoffmann, B.; da Silva, A.L.; Fortune, A.; de Lima, M.D.A.; Thomas, A. Could we expect new praziquantel derivatives? A meta pharmacometrics/pharmacoinformatics analysis of all antischistosomal praziquantel derivatives found in the literature. *SAR QSAR Environ. Res.* **2019**, *30*, 383–401. [CrossRef]

167. Qian, M. Enzymic Resolution Method for Preparing L-praziquantel. CN102911979A, 6 February 2013.

168. Yao, P.; Wu, Q.; Yuan, J.; Han, C.; Feng, J.; Zhu, D.; Ma, Y. Method for Preparing Beta-Alanine by Enzymatic Hydrolysis of High Concentration of Beta-Amino Propionitrile. CN104195193A, 10 December 2014.

169. Toth, C. Pregabalin: Latest safety evidence and clinical implications for the management of neuropathic pain. *Ther. Adv. Drug Saf.* **2014**, *5*, 38–56. [CrossRef] [PubMed]

170. Gerardi, M.C.; Atzeni, F.; Batticciotto, A.; Di Franco, M.; Rizzi, M.; Sarzi-Puttini, P. The safety of pregabalin in the treatment of fibromyalgia. *Expert Opin. Drug Saf.* **2016**, *15*, 1541–1548. [CrossRef] [PubMed]

171. Mohile, S.S.; Yeranda, S.G.; Lunge, S.M.; Patel, R.M.; Gugale, S.B.; Thakur, R.M.; Mokal, R.A.; Gangopadhyay, A.K.; Nightingale, P.D. An improved green process for the preparation of pregabalin from isovaleraldehyde via enzymatic enantioselective hydrolysis of alkyl 3-cyano-5-methylhexanoate. WO2014072785A2, 15 May 2014.

172. Mohile, S.S.; Yerande, S.G.; Lunge, S.M.; Patel, R.M.; Gugale, S.B.; Thakur, R.M.; Mokal, R.A.; Gangopadhyay, A.K.; Nightingale, P.D. An Improved Green Process for The Preparation of Pregabalin from Isovaleraldehyde Via Enzymatic Enantioselective Hydrolysis of Alkyl 3-Cyano-5-Methylhexanoate. IN2012MU03228A, 22 July 2016.

173. Xue, Y.; Zheng, Y.; Xu, Z.; Liu, Z.; Wang, Y.; Su, X.; Jia, D. Method for Preparation of 1-Cyano-Cyclohexyl-Acetic Acid Using Nitrilase Engineering Bacteria. CN104212850A, 17 December 2014.

174. Tardif, J.C.; Ford, I.; Tendera, M.; Bourassa, M.G.; Fox, K.; Initiative, I. Efficacy of ivabradine, a new selective I-f inhibitor, compared with atenolol in patients with chronic stable angina. *Eur. Heart J.* **2005**, *26*, 2529–2536. [CrossRef] [PubMed]

175. Ide, T.; Ohtani, K.; Higo, T.; Tanaka, M.; Kawasaki, Y.; Tsutsui, H. Ivabradine for the treatment of cardiovascular diseases. *Circ. J.* **2019**, *83*, 252–260. [CrossRef] [PubMed]

176. Pedragosa-Moreau, S.; Lefoulon, F. Process for the Enzymatic Synthesis of (7s)-3,4-Dimethoxybicyclo [4.2.0] Octa-1,3,5-Triene-7-Carboxylic Acid and Application in the Synthesis of Ivabradine And Salts Thereof. US20140242644A1, 28 August 2014.

177. Tao, J.; Tang, S.; Li, B.; Liu, G. Method for Preparing Monomethyl (R)-3-Hydroxyglutarate as Rosuvastatin Intermediate by Enzymolysis with Recombinant Nitrilase. CN103361386A, 23 October 2013.

178. Tao, J.; Tang, S.; Li, B.; Liu, G. Method for Preparing Monomethyl (R)-3-Hydroxyglutarate as Rosuvastatin Intermediate by Enzymolysis with Recombinant Nitrilase. WO2014205917A1, 31 December 2014.

179. Chen, L. Continuous Production Method of Hypolipemic Intermediate (R)-3-Hydroxyl Pentanedioic Acid Ethyl Ester. CN104372040A, 25 February 2015.

180. Xu, P.; Yu, H.; Han, S.; Li, L. Preparation of Fosamprenavir Intermediate (2R,3S)-1,2-Epoxy-3-Tert-Butoxycarbonylamino-4-Phenylbutane. CN104803954A, 29 July 2015.

181. Hu, L.; Xi, C.; Peng, D. Synthetic Method of Clopidogrel and Its Sulfate. CN107523594A, 29 December 2017.

182. Xue, P.; Shuai, H.; Ma, Y.; Zhang, L.; Li, P. Preparation of (S)-2-Chlorophenylglycine Methyl Ester Single Enantiomers by Enzymic Resolution. CN104293875A, 21 January 2015.

183. Zheng, Y.G.; Yin, H.H.; Yu, D.F.; Chen, X.; Tang, X.L.; Zhang, X.J.; Xue, Y.P.; Wang, Y.J.; Liu, Z.Q. Recent advances in biotechnological applications of alcohol dehydrogenases. *Appl. Microbiol. Biotechnol.* **2017**, *101*, 987–1001. [CrossRef]

184. Zheng, G.-W.; Ni, Y.; Xu, J.-H. Biocatalytic Processes for the Synthesis of Chiral Alcohols. In *Applied Biocatalysis: From Fundamental Science to Industrial Applications*; Wiley-VCH Verlag GmbH & Co. KGaA: Weinheim, Germany, 2016; pp. 219–250.

185. Kratzer, R.; Woodley, J.M.; Nidetzky, B. Rules for biocatalyst and reaction engineering to implement effective, NAD(P)H-dependent, whole cell bioreductions. *Biotechnol. Adv.* **2015**, *33*, 1641–1652. [CrossRef] [PubMed]

186. An, J.H.; Nie, Y.; Xu, Y. Structural insights into alcohol dehydrogenases catalyzing asymmetric reductions. *Crit. Rev. Biotechnol.* **2019**, *39*, 366–379. [CrossRef]

187. Chen, B.S.; de Souza, F.Z.R. Enzymatic synthesis of enantiopure alcohols: Current state and perspectives. *RSC Adv.* **2019**, *9*, 2102–2115. [CrossRef]

188. Uppada, V.; Bhaduri, S.; Noronha, S.B. Cofactor regeneration-an important aspect of biocatalysis. *Curr. Sci.* **2014**, *106*, 946–957.

189. Kara, S.; Schrittwieser, J.H.; Hollmann, F. *Strategies for Cofactor Regeneration in Biocatalyzed Reductions*; Blackwell Science Publ: Oxford, UK, 2014; pp. 209–238.

190. Berenguer-Murcia, A.; Fernandez-Lafuente, R. New Trends in the Recycling of NAD(P)H for the Design of Sustainable Asymmetric Reductions Catalyzed by Dehydrogenases. *Curr. Org. Chem.* **2010**, *14*, 1000–1021. [CrossRef]

191. Catapano, A.L. Ezetimibe: A selective inhibitor of cholesterol absorption. *Eur. Heart J. Suppl.* **2001**, *3*, E6–E10. [CrossRef]

192. Bruckert, E.; Giral, P.; Tellier, P. Perspectives in cholesterol-lowering therapy-The role of ezetimibe, a new selective inhibitor of intestinal cholesterol absorption. *Circulation* **2003**, *107*, 3124–3128. [CrossRef] [PubMed]

193. Kosoglou, T.; Statkevich, P.; Johnson-Levonas, A.O.; Paolini, J.F.; Bergman, A.J.; Alton, K.B. Ezetimibe-A review of its metabolism, pharmacokinetics and drug interactions. *Clin. Pharmacokinet* **2005**, *44*, 467–494. [CrossRef] [PubMed]

194. Mundorff, E.; De Vries, E. Ketoreductase Polypeptides for The Stereoselective Production of (4S)-3[(5S)-5-(4-Fluorophenyl)-5-hydroxypentanoyl]-4-phenyl-1,3-oxazolidin-2-one. WO2010025085A2, 4 March 2010.

195. Crowe, M.A.; Alvizo, O.; Behrouzian, B.; Bong, Y.K.; Collier, S.J.; Gohel, A.; Mavinahalli, J.; Modukuru, N.; Mundorff, E.; Smith, D.; et al. Genetically Engineered Carbonyl Reductase for Ezetimibe Synthesis. WO2011140219A1, 10 November 2011.

196. Luo, Y.; Ding, S.; Qu, X.; Sun, C. Intermediates Used for Synthesis of Ezetimibe, and Their Preparation Method and Application. CN104860980A, 26 August 2015.

197. Feng, W.; Wang, J. Preparation of Immobilized Ketoreductase or Carbonyl Reductase for Production of Ezetimibe Intermediate with Higher Efficiency and Reduced Wastes. CN105754983A, 13 July 2016.

198. Wu, J.; Chen, S.; He, X.; Yang, L.; Xu, G. Engineering Bacterium and Method for Preparation of Tert-butyl (3r,5s)-6-chloro-3,5-dihydroxy-hexanoate. CN104087546A, 8 October 2014.

199. De Lucchi, O.; Tartaggia, S.; Ferrari, C.; Galvagni, M.; Pontini, M.; Fogal, S.; Motterle, R.; Moreno, R.M.; Comely, A. Process for Preparation of Intermediates for the Synthesis of Statins. WO2014128022A1, 28 August 2014.

200. Hong, H.; Chen, C.; Li, J.; Shen, L.; Guo, L.; Tian, H. Process for Preparation of Chiral Intermediates for Statin Drugs. WO2015168998A1, 12 November 2015.

201. Scott, J.P.; Peters-Golden, M. Antileukotriene agents for the treatment of lung disease. *Am. J. Respir. Crit. Care Med.* **2013**, *188*, 538–544. [CrossRef] [PubMed]

202. Matsuse, H.; Kohno, S. Leukotriene receptor antagonists pranlukast and montelukast for treating asthma. *Expert Opin. Pharmacother.* **2014**, *15*, 353–363. [CrossRef] [PubMed]

203. Aslani, A.; Beigi, M. Design, formulation, and physicochemical evaluation of montelukast orally disintegrating tablet. *Int. J. Prev. Med.* **2016**, *7*, 8. [CrossRef]

204. Song, J.; Zhang, C.; Long, Z.; Liu, X.; Cai, S. Process for Preparation of Montelukast Sodium Intermediate. CN103936671A, 23 July 2014.

205. Lou, X.; Wang, X. Method for Preparation of Montelukast Sodium Intermediate. CN104326976A, 4 February 2015.

206. Lattanzi, S.; Brigo, F.; Cagnetti, C.; Verrotti, A.; Zaccara, G.; Silvestrini, M. Eslicarbazepine acetate in the treatment of adults with partial-onset epilepsy: An evidence-based review of efficacy, safety and place in therapy. *Core Evid.* **2018**, *13*, 21–31. [CrossRef] [PubMed]

207. Gohel, A.; Smith, D.; Wong, B.; Sukumaran, J.; Yeo, W.L.; Collier, S.J.; Novick, S. Biocatalytic Process for Preparing Eslicarbazepine and Analogs Thereof. US20140199735A1, 17 July 2014.

208. Schena, G.; Caplan, M.J. Everything You Always Wanted to Know about (3)-AR * (* But Were Afraid to Ask). *Cells* **2019**, *8*, 357. [CrossRef]

209. Morriello, G.J.; Wendt, H.R.; Bansal, A.; Di Salvo, J.; Feighner, S.; He, J.F.; Hurley, A.L.; Hreniuk, D.L.; Salituro, G.M.; Reddy, M.V.; et al. Design of a novel pyrrolidine scaffold utilized in the discovery of potent and selective human beta(3) adrenergic receptor agonists. *Bioorg. Med. Chem. Lett.* **2011**, *21*, 1865–1870. [CrossRef]

210. Chung, J.Y.L.; Campos, K.; Cleator, E.; Dunn, R.F.; Gibson, A.; Hoerrner, R.S.; Keen, S.; Lieberman, D.; Liu, Z.; Lynch, J.; et al. Process for Making Beta 3 Agonists and Intermediates. US20140242645A1, 28 August 2014.

211. Xu, F.; Chung, J.Y.L.; Moore, J.C.; Liu, Z.Q.; Yoshikawa, N.; Hoerrner, R.S.; Lee, J.; Royzen, M.; Cleator, E.; Gibson, A.G.; et al. Asymmetric synthesis of cis-2,5-disubstituted pyrrolidine, the core scaffold of beta (3)-AR agonists. *Org. Lett.* **2013**, *15*, 1342–1345. [CrossRef]

212. Krug, A.W.; Ziegler, C.G.; Bornstein, S.R. DHEA and DHEA-S, and their functions in the brain and adrenal medulla. In *Neuroactive Steroids in Brain Function, Behavior and Neuropsychiatric Disorders: Novel Strategies for Research and Treatment*; Ritsner, M.S., Weizman, A., Eds.; Springer: Dordrecht, The Netherlands, 2008; pp. 227–239.

213. Fryszkowska, A.; Quirmbach, M.S.; Gorantla, S.S.C.; Alieti, S.R.; Poreddy, S.R.; Dinne, N.K.R.; Timmanna, U.; Dahanukar, V. Processes for the Preparation of Dehydroepiandrosterone and Its Intermediates. WO2014188353A2, 27 November 2014.

214. Xie, X.; Huang, X.; Zhang, J.; Zhang, R. Biological Preparation Method of Dehydroepiandrosterone. CN105695551A, 22 June 2016.

215. Tao, J.; Le, Y. A Biological Preparation Method of (1R, 2S)-N-pyrrolidinyl Norephedrine. CN104805148A, 29 July 2015.

216. Yasuda, N.; Tan, L. Efavirenz, a non-nucleoside reverse transcriptase inhibitor (NNRTI), and a previous structurally related development candidate. In *Art of Process Chemistry*; Yasuda, N., Ed.; Wiley-VCH Verlag GmbH & Co. KGaA: Weinheim, Germany, 2011; pp. 1–43.

217. Costa, C.C.P.; Boechat, N.; da Silva, F.C.; Rosario, S.L.; Bezerra, T.C.; Bastos, M.M. The Efavirenz: Structure-activity relantionship and synthesis methods. *Rev. Virtual Quim.* **2015**, *7*, 1347–1370. [CrossRef]

218. Jin, Y.; Yao, Y.; Han, G. Enzymatic Method for Preparation of (S)-3-Piperidinol and Its Derivatives with High Chiral Purity from N-3-Piperidone and Its Derivatives. CN103898178A, 2 July 2014.

219. Wu, Z.; Ren, Z.; Liu, Y. Preparation of Thienyl Chiral Alcohol Compound as Intermediate for Synthesis of Duloxetine by Biocatalysis in the Presence of Carbonyl Reductase. CN104830924A, 12 August 2015.

220. Carter, N.J.; McCormack, P.L. Duloxetine. *CNS Drugs* **2009**, *23*, 523–541. [CrossRef] [PubMed]

221. Detke, M.J.; Lu, Y.L.; Goldstein, D.J.; Hayes, J.R.; Demitrack, M.A. Duloxetine, 60 mg once daily, for major depressive disorder: A randomized double-blind placebo-controlled trial. *J. Clin. Psychiatry* **2002**, *63*, 308–315. [CrossRef] [PubMed]

222. De Donatis, D.; Florio, V.; Forceili, S.; Saria, A.; Mercolini, L.; Serretti, A.; Conca, A. Duloxetine plasma level and antidepressant response. *Prog. Neuro-Psychopharmacol. Biol. Psychiatry* **2019**, *92*, 127–132. [CrossRef] [PubMed]

223. Savile, C.; Gruber, J.M.; Mundorff, E.; Huisman, G.W.; Collier, S.J. Ketoreductase Polypeptides for the Stereospecific Production of (S)-3-aryl-3-hydroxypropanamines from 3-aryl-3-ketopropanamines. WO2010025238A2, 4 March 2010.

224. Nie, Y.; Xu, Y.; Wang, Y. Method for Asymmetric Synthesis of Duloxetine Intermediate with Carbonyl Reductase. CN105803013A, 27 July 2016.

225. Zhang, Z.J.; Pan, J.; Ma, B.D.; Xu, J.H. Efficient biocatalytic synthesis of chiral chemicals. In *Bioreactor Engineering Research and Industrial Applications*; Springer Science and Business Media Deutschland GmbH: Berlin, Germany, 2016; Volume 155, pp. 55–106.

226. Xie, X.; Huang, X.; Zhang, J.; Zhang, R. Method for Preparing Ethyl (R)-2-hydroxy-4-phenylbutyrate. CN105732373A, 6 July 2016.

227. Lin, J.; Chestakova, A.; Gu, W.; Iding, H.; Li, J.; Linghu, X.; Meier, P.; Sha, C.; Stults, J.; Wang, Y.; et al. Process for the Manufacturing of a Pyrazolylaminopyrimidine Derivative. WO2015154674A1, 15 October 2015.

228. Blake, J.F.; Burkard, M.; Chan, J.; Chen, H.F.; Chou, K.J.; Diaz, D.; Dudley, D.A.; Gaudino, J.J.; Gould, S.E.; Grina, J.; et al. Discovery of (S)-1-(1-(4-Chloro-3-fluorophenyl)-2-hydroxyethyl)-4-(2-((1-methyl-1H-py razol-5-yl)amino)pyrimidin-4-yl)pyridin-2(1H)-one (GDC-0994), an extracellular signal-regulated kinase 1/2 (ERK1/2) inhibitor in early clinical development. *J. Med. Chem.* **2016**, *59*, 5650–5660. [CrossRef] [PubMed]

229. Zhang, M.K.; Peyear, T.; Patmanidis, I.; Greathouse, D.V.; Marrink, S.J.; Andersen, O.S.; Ingolfsson, H.I. Fluorinated Alcohols' Effects on Lipid Bilayer Properties. *Biophys. J.* **2018**, *115*, 679–689. [CrossRef]

230. Tao, J.; Zhang, W.; Liang, X. Preparation of (S)-1,1,1-Trifluoroisopropanol by Enzymic Resolution. CN104894169A, 9 Septenber 2015.

231. Bong, Y.K.; Vogel, M.; Collier, S.J.; Mitchell, V.; Mavinahalli, J. Polypeptides for a Ketoreductase-Mediated Stereoselective Route to Alpha Chloroalcohols. US8796002B2, 5 August 2014.

232. Mangas-Sanchez, J.; France, S.P.; Montgomery, S.L.; Aleku, G.A.; Man, H.; Sharma, M.; Ramsden, J.I.; Grogan, G.; Turner, N.J. Imine reductases (IREDs). *Curr. Opin. Chem. Biol.* **2017**, *37*, 19–25. [CrossRef]

233. Grogan, G.; Turner, N.J. InspIRED by nature: NADPH-dependent imine reductases (IREDs) as catalysts for the preparation of chiral amines. *Chem. Eur. J.* **2016**, *22*, 1900–1907. [CrossRef]

234. Maugeri, Z.; Rother, D. Application of Imine Reductases (IREDs) in Micro-Aqueous Reaction Systems. *Adv. Synth. Catal.* **2016**, *358*, 2745–2750. [CrossRef]

235. Maugeri, Z.; Rother, D. Reductive amination of ketones catalyzed by whole cell biocatalysts containing imine reductases (IREDs). *J. Biotechnol.* **2017**, *258*, 167–170. [CrossRef]

236. Lenz, M.; Borlinghaus, N.; Weinmann, L.; Nestl, B.M. Recent advances in imine reductase-catalyzed reactions. *World, J. Microbiol. Biotechnol.* **2017**, *33*, 10. [CrossRef] [PubMed]

237. Cosgrove, S.C.; Brzezniak, A.; France, S.P.; Ramsden, J.I.; Mangas-Sanchez, J.; Montgomery, S.L.; Heath, R.S.; Turner, N.J. Imine reductases, reductive aminases, and amine oxidases for the synthesis of chiral amines: Discovery, characterization, and synthetic applications. In *Enzymes in Synthetic Biology*; Scrutton, N., Ed.; Elsevier Academic Press Inc.: San Diego, CA, USA, 2018; Volume 608, pp. 131–149.

238. Gamenara, D.; Domínguez de María, P. Enantioselective imine reduction catalyzed by imine reductases and artificial metalloenzymes. *Org. Biomol. Chem.* **2014**, *12*, 2989–2992. [CrossRef] [PubMed]

239. Velikogne, S.; Resch, V.; Dertnig, C.; Schrittwieser, J.H.; Kroutil, W. Sequence-based in-silico discovery, characterisation, and biocatalytic application of a set of imine reductases. *ChemCatChem* **2018**, *10*, 3236–3246. [CrossRef] [PubMed]

240. Patil, M.D.; Grogan, G.; Bommarius, A.; Yun, H. Oxidoreductase-catalyzed synthesis of chiral amines. *ACS Catal.* **2018**, *8*, 10985–11015. [CrossRef]

241. Agard, N.J.; Alvizo, O.; Mayo, M.A.; Minor, S.N.; Riggins, J.N.; Moore, J.C. Variants of the Opine Dehydrogenase of Arthrobacter C1 for Use in the Reductive Amination of Ketones and Amines. US20150132807A1, 14 May 2015.

242. Chen, H.; Moore, J.; Collier, S.J.; Smith, D.; Nazor, J.; Hughes, G.; Janey, J.; Huisman, G.; Novick, S.; Agard, N.; et al. Engineered Imine Reductases and Methods for the Reductive Amination of Ketone and Amine Compounds. WO2013170050A1, 14 November 2013.

243. Debarge, S.; Erdman, D.T.; O'Neill, P.M.; Kumar, R.; Karmilowicz, M.J. Process and Intermediates for the Preparation of Pregabalin. WO2014155291A1, 2 October 2014.

244. Turner, N.J. Enantioselective oxidation of C-O and C-N bonds using oxidases. *Chem. Rev.* **2011**, *111*, 4073–4087. [CrossRef] [PubMed]

245. Hollmann, F.; Arends, I.; Buehler, K.; Schallmey, A.; Buhler, B. Enzyme-mediated oxidations for the chemist. *Green Chem.* **2011**, *13*, 226–265. [CrossRef]

246. Romano, D.; Villa, R.; Molinari, F. Preparative biotransformations: Oxidation of alcohols. *ChemCatChem* **2012**, *4*, 739–749. [CrossRef]

247. Wu, S.K.; Liu, J.; Li, Z. Organic synthesis via oxidative cascade biocatalysis. *Synlett* **2016**, *27*, 2644–2658.

248. Dong, J.J.; Fernandez-Fueyo, E.; Hollmann, F.; Paul, C.E.; Pesic, M.; Schmidt, S.; Wang, Y.H.; Younes, S.; Zhang, W.Y. Biocatalytic oxidation reactions: A chemist's perspective. *Angew. Chem.Int. Ed.* **2018**, *57*, 9238–9261. [CrossRef]

249. Liu, J.; Wu, S.K.; Li, Z. Recent advances in enzymatic oxidation of alcohols. *Curr. Opin. Chem. Biol.* **2018**, *43*, 77–86. [CrossRef]

250. Schulz, S.; Girhard, M.; Urlacher, V.B. Biocatalysis: Key to Selective Oxidations. *ChemCatChem* **2012**, *4*, 1889–1895. [CrossRef]

251. Gamenara, D.; Seoane, G.; Méndez, P.S.; Domínguez de de María, P. *Redox Biocatalysis: Fundamentals and Applications*; John Wiley & Sons, Inc.: Hoboken, NJ, USA, 2012.

252. Wang, Y.H.; Lan, D.M.; Durrani, R.; Hollmann, F. Peroxygenases en route to becoming dream catalysts. What are the opportunities and challenges? *Curr. Opin. Chem. Biol.* **2017**, *37*, 1–9. [CrossRef] [PubMed]

253. Van Rantwijk, F.; Sheldon, R.A. Selective oxygen transfer catalysed by heme peroxidases: Synthetic and mechanistic aspects. *Curr. Opin. Biotechnol.* **2000**, *11*, 554–564. [CrossRef]

254. Hofrichter, M.; Ullrich, R. Oxidations catalyzed by fungal peroxygenases. *Curr. Opin. Chem. Biol.* **2014**, *19*, 116–125. [CrossRef] [PubMed]

255. Urlacher, V.B.; Girhard, M. Cytochrome P450 monooxygenases: An update on perspectives for synthetic application. *Trends Biotechnol.* **2012**, *30*, 26–36. [CrossRef] [PubMed]

256. Fasan, R. Tuning P450 enzymes as oxidation catalysts. *ACS Catal.* **2012**, *2*, 647–666. [CrossRef]

257. Bucko, M.; Gemeiner, P.; Schenkmayerova, A.; Krajcovic, T.; Rudroff, F.; Mihovilovic, M.D. Baeyer-Villiger oxidations: Biotechnological approach. *Appl. Microbiol. Biotechnol.* **2016**, *100*, 6585–6599. [CrossRef] [PubMed]

258. Bondani, P.B.; Fraaije, M.W.; de Gonzalo, G. Recent developments in flavin-based catalysis: Enzymatic sulfoxidations. In *Green Biocatalysis*; Patel, R.N., Ed.; John Wiley & Sons, Inc.: Hoboken, NJ, USA, 2016; pp. 149–164.

259. Urlacher, V.B.; Girhard, M. Cytochrome P450 monooxygenases in biotechnology and synthetic biology. *Trends Biotechnol.* **2019**, *37*, 882–897. [CrossRef]

260. Strand, D.S.; Kim, D.; Peura, D.A. 25 Years of proton pump inhibitors: A comprehensive review. *Gut Liver* **2017**, *11*, 27–37. [CrossRef]

261. Bong, Y.K.; Clay, M.D.; Collier, S.J.; Mijts, B.; Vogel, M.; Zhang, X.; Zhu, J.; Nazor, J.; Smith, D.; Song, S. Synthesis of Prazole Compounds. US8895271B2, 25 November 2014.

262. Kayser, M.M.; Clouthier, C.M. New bioorganic reagents: Evolved cyclohexanone monooxygenases—Why is it more selective? *J. Org. Chem.* **2006**, *71*, 8424–8430. [CrossRef]

263. Ang, E.L.; Clay, M.D.; Behrouzian, B.; Eberhard, E.; Collier, S.J.; Fu Fan, J.; Smith, D.; Song, S.; Alvizo, O.; Widegren, M.; et al. Chemoenzymic Synthesis of Armodafinil Using Genetically Modified Acinetobacter Cyclohexanone Monooxygenase. WO2012078800A2, 14 June 2012.

264. Phillips, J.B.; Simmons, R.G.; Arnold, R.D. A single dose of armodafinil significantly promotes vigilance 11 h post-dose. *Mil. Med.* **2011**, *176*, 833–839. [CrossRef] [PubMed]

265. Garnock-Jones, K.P.; Dhillon, S.; Scott, L.J. Armodafinil. *CNS Drugs* **2009**, *23*, 793–803. [CrossRef] [PubMed]

266. Kino, K.; Furuya, T. Production of (−)-rotundone from α-guaiene catalyzed by cytochrome P450 enzymes. EP3255151A2, 13 December 2017.

267. Wood, C.; Siebert, T.E.; Parker, M.; Capone, D.L.; Elsey, G.M.; Pollnitz, A.P.; Eggers, M.; Meier, M.; Vossing, T.; Widder, S.; et al. From wine to pepper: Rotundone, an obscure sesquiterpene, is a potent spicy aroma compound. *J. Agric. Food Chem.* **2008**, *56*, 3738–3744. [CrossRef] [PubMed]

268. Pecyna, M.J.; Schnorr, K.M.; Ullrich, R.; Scheibner, K.; Kluge, M.G.; Hofrichter, M. Fungal Peroxygenases and Use for Regioselective Oxidation of N-Heterocycles. WO2008119780A2, 9 October 2008.

269. Steffen-Munsberg, F.; Vickers, C.; Kohls, H.; Land, H.; Mallin, H.; Nobili, A.; Skalden, L.; van den Bergh, T.; Joosten, H.J.; Berglund, P.; et al. Bioinformatic analysis of a PLP-dependent enzyme superfamily suitable for biocatalytic applications. *Biotechnol. Adv.* **2015**, *33*, 566–604. [CrossRef] [PubMed]

270. Patil, M.D.; Grogan, G.; Bommarius, A.; Yun, H. Recent Advances in omega-transaminase-mediated biocatalysis for the enantioselective synthesis of chiral amines. *Catalysts* **2018**, *8*, 254. [CrossRef]

271. Bornscheuer, U.; Hanlon, S.P.; Iding, H.; Pavlidis, I.; Spurr, P.; Steffen Weiss, M.; Wirz, B. Variants of a Bacterial Transaminase with Increased Catalytic Activity for the Synthesis of Chiral Amines. WO2016166120A1, 20 October 2016.

272. Dhawan, I.K.; Miller, G.; Zhang, X. Transaminase Polypeptides with Improved Thermostability and Enantioselectivity for the Synthesis of Chiral Amines. WO2010081053A2, 15 July 2010.
273. Hughes, G.; Devine, P.N.; Fleitz, F.J.; Grau, B.T.; Limanto, J.; Savile, C.; Mundorff, E. Engineering of Arthrobacter Transaminase Variants, and Use for Biocatalytic Asymmetric Preparation of Chiral Amines from Prochiral Ketones. WO2011005477A1, 13 January 2011.
274. Savile, C.; Mundorff, E.; Moore, J.C.; Devine, P.N.; Janey, J.M. Construction of Arthrobacter KNK168 Transaminase Variants for Biocatalytic Manufacture of Sitagliptin. WO2010099501A2, 2 September 2010.
275. Schuermann, M.; Peeters, W.P.H.; Smeets, N.H.J.; Schwab, H.; Steiner, K.; Lypetska, K.M.; Strohmeier, G. Stereoselective Transamination Catalyzed by Novel R-Selective Transaminases. WO2012007548A1, 19 January 2012.
276. Hong, H.; Gao, F.; Li, Y.; Zhang, Y.; Li, S. Sequence of Transaminase Mutants and Its Use for Synthesizing R configuration Chiral Amine. CN104328094A, 4 February 2015.
277. Tao, R.; Zhu, F.; Shen, Q.; Sun, L.; Shen, Z.; Jiang, Y.; Yang, C. Transaminase for Producing Antiepileptic Drug Levetiracetam Intermediate L-2-aminobutyric acid. CN105441403A, 30 March 2016.
278. Wang, H.; Xie, Y.; Wang, J.; Fan, H.; Wei, D. S type ω-transaminase ATA-W12 and Gene and Application Thereof. CN106520719A, 22 March 2017.
279. Shin, J.-S.; Park, E.-S.; Han, S.-W. Omega-transaminase Bacterial Mutants with Activity Improvements Toward Ketones and Methods for Producing Optically Pure Amines. US20160298092A1, 13 October 2016.
280. Shin, J.S.; Park, E.S. Production of Optically Active Amines by Using Omega Transaminase Derived from Ochrobactrum Anthropi and Devoid of Substrate and Product Inhibition. KR2014123166A, 22 October 2014.
281. Truppo, M.D.; Janey, J.M.; Hughes, G. Immobilization of Transaminases on Resin Carriers for Use in the Manufacture of Sitagliptin. WO2012177527A1, 27 December 2012.
282. Huang, Q.; Zhang, X.; Xiang, S. One Kind of Immobilization Method of Transaminase. CN104774830A, 15 July 2015.
283. Truppo, M.D.; Journet, M.; Strotman, H.; McMullen, J.P.; Grosser, S.T. Immobilized Transaminases for the Preparation of Sitagliptin by Stereospecific Amination of Prositagliptin Ketone. WO2014133928A1, 4 September 2014.
284. Aroda, V.R.; Henry, R.R.; Han, J.; Huang, W.Y.; DeYoung, M.B.; Darsow, T.; Hoogwerf, B.J. Efficacy of GLP-1 Receptor Agonists and DPP-4 Inhibitors: Meta-Analysis and Systematic Review. *Clin. Ther.* **2012**, *34*, 1247–1258. [CrossRef]
285. Luo, Y.; Ding, S.; Qu, X.; Guo, F. Immobilized Transaminase and Its Application in Synthesizing Sitagliptin Intermediate. CN104805069A, 29 July 2015.
286. Luo, Y.; Ding, S.; Qu, X. Immobilized Transaminase and Its Application in Synthesis of Sitagliptin Intermediate. CN105219745A, 6 January 2016.
287. Luo, Y.; Ding, S.; Qu, X. Mycobacterium Vanbaalenii Transaminase and Its Application in Preparing Sitagliptin Intermediate. CN105018440A, 4 November 2015.
288. Savile, C.K.; Janey, J.M.; Mundorff, E.C.; Moore, J.C.; Tam, S.; Jarvis, W.R.; Colbeck, J.C.; Krebber, A.; Fleitz, F.J.; Brands, J.; et al. Biocatalytic asymmetric synthesis of chiral amines from ketones applied to sitagliptin manufacture. *Science* **2010**, *329*, 305–309. [CrossRef]
289. Robins, K.; Bornscheuer, U.; Hoehne, M. Process for Preparation of Optically Active N-protected 3-Aminopyrrolidine or Optically Active N-protected 3-aminopiperidine and the Corresponding Ketones by Optical Resolution of the Racemic Amine Mixtures Employing a Bacterial Omega-Transaminase. WO2008028654A1, 13 March 2008.
290. Robins, K.; Bornscheuer, U.; Hoehne, M. Preparation of Optically Active Chiral Amines Using a Stereoselective transaminase. EP1818411A1, 15 August 2007.
291. Lin, Y.; Hong, H.; Yan, J.; Wang, Z. Preparation of 3-Aminopyrrolidine Compounds and Their Application. CN104262225A, 7 January 2015.
292. Stass, H.; Dalhoff, A.; Kubitza, D.; Schuhly, U. Pharmacokinetics, safety, and tolerability of ascending single doses of moxifloxacin, a new 8-methoxy quinolone, administered to healthy subjects. *Antimicrob. Agents Chemother.* **1998**, *42*, 2060–2065. [CrossRef]
293. Limanto, J.; Beutner, G.; Grau, B.; Janey, J.; Klapars, A.; Ashley, E.R.; Strotman, H.R.; Truppo, M.D.; Hughes, G.; Cabirol, F.; et al. Biocatalysts and Methods for the Synthesis of (1R,2R)-2-(3,4-dimethoxyphenethoxy)cyclohexanamine. US8932836B2, 13 January 2015.

294. Kossaify, A. Vernakalant in atrial fibrillation: A relatively new weapon in the armamentarium against an old enemy. *Drug Target. Insights* **2019**, *13*, 7. [CrossRef]

295. Hall, A.J.M.; Mitchell, A.R.J. Introducing Vernakalant into clinical practice. *Arrhythm. Electrophysiol. Rev.* **2019**, *8*, 70–74. [CrossRef]

296. Cabirol, F.; Gohel, A.; Oh, S.H.; Smith, D.; Wong, B.; LaLonde, J. Biocatalysts and Methods for the Synthesis of (S)-3-(1-aminoethyl)-phenol. WO2011159910A2, 22 December 2011.

297. Cabirol, F.; Chen, H.; Gohel, A.; Salim, P.; Smith, D.; Janey, J.; Kosjek, B.; Tang, W.L.; Hsieh, H.; Pham, S. Biocatalysts and Methods for the Synthesis of Substituted Lactams. US9109209B2, 18 August 2015.

298. Limanto, J.; Beutner, G.L.; Yin, J.; Klapars, A.; Ashley, E.R.; Strotman, H.R.; Truppo, M.D.; Chung, C.K.; Hughes, G.; Liu, Z.; et al. Process for Preparing Aminocyclohexyl Ether Compounds. US9006460B2, 14 April 2015.

299. Austad, B.; Bahadoor, A.; Belani, J.D.; Janardanannair, S.; Johannes, C.W.; Keaney, G.F.; Lo, C.K.; Wallerstein, S.L. Enzymatic Transamination of Cyclopamine Analogs. WO2011017551A1, 10 February 2011.

300. Crowe, M.; Foulkes, M.; Francese, G.; Grimler, D.; Kuesters, E.; Laumen, K.; Li, Y.; Lin, C.; Nazor, J.; Ruch, T.; et al. Chemical Process for Preparing Spiroindolones and Intermediates Thereof. WO2013139987A1, 26 September 2013.

301. Sieber, V.; Grammann, K.; Ruehmann, B.; Haas, T.; Pfeffer, J.; Doderer, K.; Rollmann, C.; Skerra, A.; Rausch, C.; Lerchner, A. Enzymic transamination of multicyclic dianhydro diuloses. WO2010089171A2, 12 August 2010.

302. Tao, J.; Liang, X.; Le, Y. Method for Preparation of (S)-2-amino-1-butanol by Biological Process. CN105177071A, 23 December 2015.

303. Qu, D.; Huang, Y. Synthesis Method of Efinaconazole Intermediate. CN105039450A, 11 November 2015.

304. Tatsumi, Y.; Nagashima, M.; Shibanushi, T.; Iwata, A.; Kangawa, Y.; Inui, F.; Siu, W.J.J.; Pillai, R.; Nishiyama, Y. Mechanism of action of efinaconazole, a novel triazole antifungal agent. *Antimicrob. Agents Chemother.* **2013**, *57*, 2405–2409. [CrossRef] [PubMed]

305. Gupta, A.K.; Cernea, M. How effective is efinaconazole in the management of onychomycosis? *Expert Opin. Pharmacother.* **2016**, *17*, 611–618. [CrossRef] [PubMed]

306. Occhipinti, A.; Berlicki, L.; Giberti, S.; Dziedziola, G.; Kafarski, P.; Forlani, G. Effectiveness and mode of action of phosphonate inhibitors of plant glutamine synthetase. *Pest Manag. Sci.* **2010**, *66*, 51–58. [CrossRef] [PubMed]

307. Yang, L.; Zhou, H.; Meng, L.; Liu, S.; Wei, Y.; Gu, S.; Yuan, X.; Fang, H.; Liu, Y.; Xu, G.; et al. A Production Method of L-glufosinate ammonium. CN105603015A, 25 May 2016.

308. Green, B.M.; Gradley, M.L. Two-step Enzymic Synthesis of L-glufosinate from Racemic Mixture. US20170253897A1, 7 September 2017.

309. Bulger, P.G.; Kosjek, B.; Rivera, N. Biocatalytic Transamination Process for the Enantioselective Syntheses of 3-Arylpiperidine, Key Intermediate in the Prepn. of Poly (adp-ribose) Polymerase Inhibitors. WO2014088984A1, 12 June 2014.

310. Kim, M.H.; Han, G.S. Method for Producing D-alanine Using Fusion Enzyme. KR2017132446A, 4 December 2017.

311. Yun, D.H.; Lee, H.S. Enzymatic Production Method of Optically Active Beta-Amino Acids Including Intermediate for the Synthesis of Sitagliptin. KR101565439B1, 18 November 2015.

312. Shin, J.S.; Park, E.S.; Dong, J.Y. Simultaneous Production of High-Purity Optically Active Alkylamine and Aryl Alkylamine Using Transaminase. KR2014108422A, 11 September 2014.

313. Shin, J.-S.; Park, E.-S.; Malik, M.S. Method for Preparing Optically Active Amine Compound Using Deracemization. WO2014092496A1, 19 June 2014.

314. Huisman, G.W.; Agard, N.J.; Mijts, B.; Vroom, J.; Zhang, X. Robust Variants of a Phenylalanine Ammonia Lyase with Increased Catalytic Activity for Therapeutic Use. WO2014172541A2, 23 October 2014.

315. Wu, B.; Szymanski, W.; Crismaru, C.G.; Feringa, B.L.; Janssen, D.B. C-N Lyases Catalyzing Addition of Ammonia, Amines, and Amides to C=C and C=O Bonds. In *Enzyme Catalysis in Organic Synthesis*, 3rd ed.; Drauz, K., Groeger, H., May, O., Eds.; Jonh Wiley and Sons, Inc.: Hoboken, NJ, USA, 2012; pp. 749–778.

316. Huisman, G.W.; Agard, N.J.; Elgart, D.; Zhang, X. Tyrosine: Ammonia Lyase Variants Stable at Acid pH and Resistant to Proteolysis and Their Use in Treatment of Metabolic Disease. WO2015161019A1, 22 October 2015.

317. Weiner, D.; Varvak, A.; Richardson, T.; Podar, M.; Burke, E.; Healey, S. Lyase Enzymes, Nucleic Acids Encoding Them and Methods for Their Production and Commercial Use. WO2006099207A2, 21 September 2006.

318. Bruehlmann, F.; Fourage, L.; Wahler, D. Engineering of Modified Guava 13-hydroperoxide Lyase Variants with Improved Enzymic Activity for Use in the Production of Aldehydes and Alcohols. WO2009001304A1, 31 December 2008.

319. Asano, Y.; Dadashipour Lakmehsar, M.; Ishida, Y. Purification and characterization of novel (R)-hydroxynitrile lyase. WO2015133462A1, 11 September 2015.

320. Eggert, T.; Andexer, J. An (R)-hydroxynitrile Lyase from a Non-Cyanogenic Member of the Brassicaceae for Use in Stereoselective Addition. WO2008067807A2, 12 June 2008.

321. Glieder, A. R-Hydroxynitrile Lyase Random Variants and Their Use for Preparing Optically Pure, Sterically Hindered Cyanohydrins. WO2008071695A1, 19 June 2008.

322. Xiao, Y.; Yang, W.; Chen, M.; Gong, B.; Yan, Y.; Tan, C. Levodopa Crystal Powder and Preparation Method Thereof. CN106117072A, 16 November 2016.

323. De Deurwaerdère, P.; Di Giovanni, G.; Millan, M.J. Expanding the repertoire of L-DOPA's actions: A comprehensive review of its functional neurochemistry. *Prog. Neurobiol.* **2017**, *151*, 57–100. [CrossRef] [PubMed]

324. Jenner, P. Molecular mechanisms of L-DOPA-induced dyskinesia. *Nat. Rev. Neurosci.* **2008**, *9*, 665–677. [CrossRef] [PubMed]

325. Cools, R. Dopaminergic modulation of cognitive function-implications for L-DOPA treatment in Parkinson's disease. *Neurosci. Biobehav. Rev.* **2006**, *30*, 1–23. [CrossRef] [PubMed]

326. Rother, D.; Pohl, M.; Sehl, T.; Baraibar, A.G. Enzymic Twostep Process for the Production of Cathine. DE102013009631A1, 11 December 2014.

327. Rother, D.; Pohl, M.; Sehl, T.; Baraibar, A.G. Process for Producing Cathine. WO2014198247A1, 18 December 2014.

328. Hauner, H.; Hastreiter, L.; Werdier, D.; Chen-Stute, A.; Scholze, J.; Bluher, M. Efficacy and safety of cathine (nor-pseudoephedrine) in the treatment of obesity: A randomized dose-finding study. *Obes. Facts* **2017**, *10*, 407–419. [CrossRef] [PubMed]

329. Alsufyani, H.A.; Docherty, J.R. Cathine and MHA act as indirect sympathomimetics to produce cardiovascular actions. *Br. J. Pharmacol.* **2019**, *176*, 3036.

330. Clapés, P. Enzymatic C-C Bond Formation. In *Organic Synthesis Using Biocatalysis*; Goswami, A., Stewart, J.D., Eds.; Academic Press: Cambridge, MA, USA, 2016; pp. 285–337.

331. Brovetto, M.; Gamenara, D.; Mendez, P.S.; Seoane, G.A. C-C Bond-forming lyases in organic synthesis. *Chem. Rev.* **2011**, *111*, 4346–4403. [CrossRef]

332. Sehl, T.; Hailes, H.C.; Ward, J.M.; Menyes, U.; Pohl, M.; Rother, D. Efficient 2-step biocatalytic strategies for the synthesis of all nor(pseudo) ephedrine isomers. *Green Chem.* **2014**, *16*, 3341–3348. [CrossRef]

333. Sehl, T.; Hailes, H.C.; Ward, J.M.; Wardenga, R.; von Lieres, E.; Offermann, H.; Westphal, R.; Pohl, M.; Rother, D. Two steps in one pot: Enzyme cascade for the synthesis of nor(pseudo)ephedrine from inexpensive starting materials. *Angew. Chem. Int. Ed.* **2013**, *52*, 6772–6775. [CrossRef] [PubMed]

334. Erdmann, V.; Sehl, T.; Frindi-Wosch, I.; Simon, R.C.; Kroutil, W.; Rother, D. Methoxamine Synthesis in a Biocatalytic 1-Pot 2-Step Cascade Approach. *ACS Catal.* **2019**, *9*, 7380–7388. [CrossRef]

335. Mrak, P.; Zohar, T.; Oslaj, M.; Kopitar, G. Enzymatic Synthesis of Statins and Intermediates Thereof. WO2012095244A2, 19 July 2012.

336. Casar, Z.; Mesar, T.; Kopitar, G.; Mrak, P.; Oslaj, M. Synthesis of Statins from (4R,6S)-6-(dialkoxymethyl) tetrahydro-2H-pyran-2,4-diol. WO2008119810A2, 9 October 2008.

337. Cluzeau, J.; Casar, Z.; Mrak, P.; Oslaj, M.; Kopitar, G. Process for the Production of ((2S,4R)-4,6-dihydroxytetrahydro-2H-pyran-2-yl)methyl Carboxylates and Its Further use to Synthesisze Statins. WO2009092702A2, 20 July 2009.

338. Bauer, D.W.; Hu, S.; O'Neill, P.M.; Watson, T.J.N. Process for Preparing Chiral Compounds Using 2-Deoxyribose-5-phosphate aldolase. WO2009019561A2, 12 February 2009.

339. Matsumoto, K.; Kazuno, Y.; Higashimura, N.; Ohshima, T.; Sakuraba, H. Enzymic Preparation of Chiral Hydroxyaldehydes. WO2005098012A1, 20 October 2005.

340. Zhu, G.; Qi, X.; Li, B.; Wang, C.; Zhang, H.; Liu, Y.; Pan, X. Chemoenzymic Synthesis of Atorvastatin Calcium Side Chain with High Chiral Purity. CN103060396A, 12 September 2013.

341. Oh, D.G.; Lee, S.H.; Hong, S.H. Method for Producing Tagatose from Fructose and Enzyme Composition Therefor. KR1480422B1, 5 February 2015.

342. Oh, D.K.; Hong, S.H.; Lee, S.H. Method for Producing Tagatose from Fructose and Enzyme Composition Therefor. WO2015016544A1, 5 February 2015.
343. Xu, G.; Zeng, H.; Zhao, S.; Chen, M. Method for Producing L-alpha-aminobutyric acid with High Optical Activity by Biological Catalysis. CN104774881A, 15 July 2015.
344. Kroutil, W.; Busto, E.; Simon, R. Method for Producing p-vinylphenols. AT516155A4, 15 March 2016.
345. Kroutil, W.; Busto, E.; Simon, R. Method for Producing p-vinylphenols. WO2016141397A1, 15 September 2016.

Review

Pseudokinases: From Allosteric Regulation of Catalytic Domains and the Formation of Macromolecular Assemblies to Emerging Drug Targets

Andrada Tomoni [1], Jonathan Lees [1], Andrés G. Santana [2], Victor M. Bolanos-Garcia [1,*] and Agatha Bastida [2,*]

[1] Department of Biological and Medical Sciences, Faculty of Health and Life Sciences, Oxford Brookes University, Oxford OX3 OBP, UK; andradatomoni@gmail.com (A.T.); jlees@brookes.ac.uk (J.L.)

[2] Departamento de Química Bio-orgánica, IQOG, c/Juan de la Cierva 3, E-28006 Madrid, Spain; andres.g.santana@csic.es

* Correspondence: vbolanos-garcia@brookes.ac.uk (V.M.B.-G.); agatha.bastida@csic.es (A.B.); Tel.: +44-01865-484146 (V.M.B.-G.); +34-9156-188-06 (A.B.)

Received: 7 August 2019; Accepted: 13 September 2019; Published: 19 September 2019

Abstract: Pseudokinases are a member of the kinase superfamily that lack one or more of the canonical residues required for catalysis. Protein pseudokinases are widely distributed across species and are present in proteins that perform a great diversity of roles in the cell. They represent approximately 10% to 40% of the kinome of a multicellular organism. In the human, the pseudokinase subfamily consists of approximately 60 unique proteins. Despite their lack of one or more of the amino acid residues typically required for the productive interaction with ATP and metal ions, which is essential for the phosphorylation of specific substrates, pseudokinases are important functional molecules that can act as dynamic scaffolds, competitors, or modulators of protein–protein interactions. Indeed, pseudokinase misfunctions occur in diverse diseases and represent a new therapeutic window for the development of innovative therapeutic approaches. In this contribution, we describe the structural features of pseudokinases that are used as the basis of their classification; analyse the interactome space of human pseudokinases and discuss their potential as suitable drug targets for the treatment of various diseases, including metabolic, neurological, autoimmune, and cell proliferation disorders.

Keywords: pseudokinases; signal transduction; cancer therapy; tyrosine/serine/threonine phosphorylation; new drug targets; interactome

1. Introduction

Pseudokinases represent approximately 10% to 40% of the kinome of a multicellular organism [1–4]. Systematic analysis of the human genome has revealed that approximately one-tenth of all protein kinases are predicted to be catalytically inactive and therefore signalling occurs through other mechanisms. Pseudokinases are not restricted to multicellular organisms [5,6] and form part of the bigger pseudoenzyme superfamily [7–9].

So far, the picture that is emerging is that pseudokinases, like their catalytically active counterparts, play pivotal roles in cellular signalling systems as mediators of protein interactions, often involving allosteric regulation of interaction partners, including bona fide protein kinases; the control of subcellular localisation and trafficking; and the assembly of larger macromolecular complexes. Moreover, pseudokinases are often dysregulated in a variety of diseases ranging from developmental, metabolic and neurological disorders to cancer and autoimmune diseases [10–15]. Although important insights into pseudokinase function remain to be established, the fact that several clinically approved kinase inhibitors pharmacologically regulate the noncatalytic functions of active kinases, suggests that similar properties may be exploited in pseudokinases associated with human malignancies [16–20].

Independent crystal structures of pseudokinases have revealed that even though they do not retain canonical motifs found in bona fide catalytically active kinases, they are able to adopt local conformational transitions that resemble those present in the former class. Features emerge the possibility of designing new small molecules that could be of therapeutic value by selectively inducing non-functional conformations in pseudokinases. At the same time, the catalytic deficiencies of pseudokinases make these proteins a powerful toolbox for the study of mechanisms of allosteric regulation [21–23].

In this contribution, we show the structural features that are used to classify pseudokinases and discuss the interactome space of human pseudokinases and the promise they represent for the treatment of human diseases, including cancer.

2. Types of Pseudokinases Proteins

Pseudokinases have the same overall kinase domain fold. Namely, an N-terminal lobe, composed of five β-sheets, that is connected to the C-terminal lobe via a flexible hinge region. The catalytic site is located at the lobes interface. However, pseudokinases are proteins that lack one or more key conserved catalytic residues present in active kinases [5,7] (Figure 1).

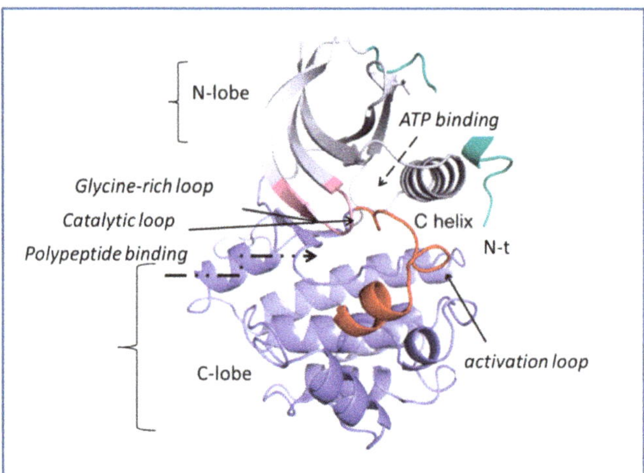

Figure 1. General structure of a kinase protein domain with key structural motifs.

This includes the lysine residue known to position ATP during phosphoryl transfer, which forms the canonical β3-lys/αC-Glu salt-bridge interaction within the VAIK (Val-Ala-Ile-Lys) motif; the aspartic acid in the catalytic loop, known as the catalytic residue within the HRD (His-Arg-Asp) motif; and the aspartic acid in the activation loop, which binds the divalent cations within the DFG (Asp-Phe-Gly) motif (Figure 2). The N-lobe is highly conserved across the kinome [24–26] whereas the C-lobe presents an open surface that facilitates protein/protein interactions [27]. Pseudokinases can adopt conformations that recapitulate features of either the "on" or "off" state of catalytically active protein kinases, and in many cases, these conformations are critical for their physiological roles.

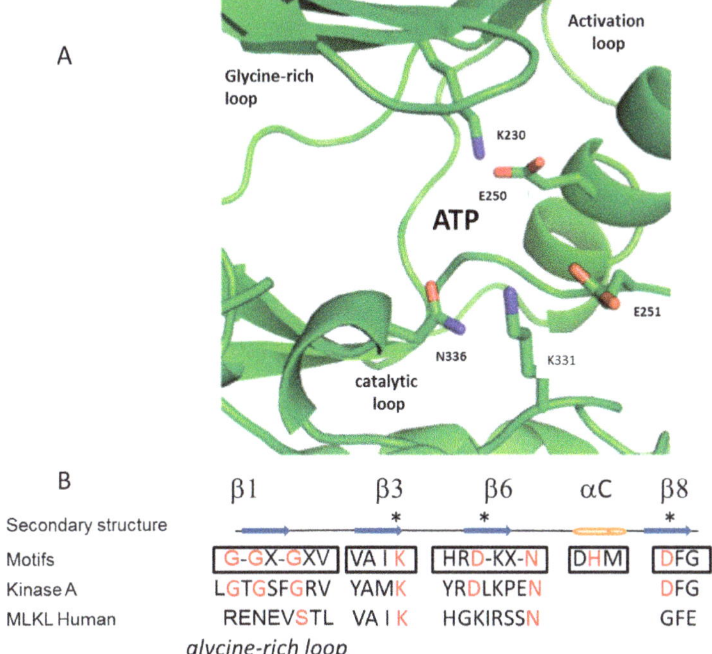

Figure 2. (**A**) the pseudoactive site of human complex ATP-MLKL pseudokinase structure (PDB:5knj), (**B**) secondary structure and amino acid sequence alignment of the catalytic motif of protein kinase A and the MLKL pseudokinase domain. (*) Residues known to mediate catalysis.

The VAIK motif is located in the β3 strand of the N-lobe, which positions the catalytic lysine to coordinate the α and β phosphates of ATP. The HRD-XX-X motif is required for catalysis. This motif is mapped onto the catalytic loop. The aspartate residue in this motif serves as the catalytic base during ATP hydrolysis, and its substitution results in a catalytically inactive kinase. Several pseudokinases carry HRD mutations or are missing this motif entirely, including HER3, integrin-linked protein kinase (ILK) and Mixed Lineage Kinase Domain-Like (MLKL). Other sequence alterations in pseudokinases include substitutions in the glycine-rich loop located in the N-lobe. The glycine-rich loop usually conforms to the consensus sequence GXGXXG in active kinases. The absence of side chains in the glycine residues allows for close contact of the glycine-rich loop with the adenosine ring of ATP, which enables nucleotide binding and proper positioning of ATP for catalysis. In pseudokinases, the glycines are often replaced by larger amino acids, frequently negatively charged, that interfere with ATP binding.

Conformational changes within the activation loop are sometimes accompanied by the motions of the N-lobe located DFG motif. In the active conformation, the aspartate points into the active site and coordinates a Mg^{2+} ion that interacts with the β-phosphate and γ-phosphate of the ATP molecule. In the inactive conformational state, the aspartate rotates ~180° away from the active site. Many pseudokinases lack this motif entirely while others carry mutations in their DFG motifs.

Pseudokinases can be classified on the basis of whether or not they can retain the capacity to interact with nucleotides and/or divalent cations [28,29]. The key structural feature of such classification is discussed as follows:

(a) pseudokinases do not bind nucleotides or cations. The pseudokinases Trib1 (pdb: 5cek, 5cem, 6dc0), VRK3 (pdb: 2jll), ROR2 (pdb: 3zzw, 4gt4), Pragmin (pdb: 5ve6, 6ewx) and MviN (pdb: 3otv) proteins belong to this group. The crystal structures of these proteins present highly distorted nucleotide binding sites that indicate the absence of a well-defined catalytic pocket. Nevertheless, the pseudo-active site could potentially be able to accommodate non-conventional ligands [30]. In the human, three Tribbles (TRIB) homologs have been identified: Trib1, Trib2 and Trib3 [31–33]. These pseudokinases play important functions in lipoprotein metabolism, immune response and cellular differentiation and proliferation and are organised as an N-terminal PEST region, a pseudokinase domain with an atypical DFG metal-binding motif and a C-terminal binding region.

(b) Pseudokinases bind nucleotides in the absence of cations. This group includes the pseudokinases Stradα (pdb: 2wtk, 3gni), MLKL (pdb: 4btf, 4m67, 4m68, 4m69, 4mwi, 5knj, 5ko1), FAM20A (pdb: 5wrr, 5wrs, 5yh2, 5yh3), CASK (pdb: 3c0g, 3c0H, 3c0I, 3mrf, 3mfs, 3tac), Ulk4 and Trib2/3 [34–36]. With the exception of FAM20A, this class of pseudokinases lacks the aspartate residue found in the DFG motifs of active kinases (Figure 2). For example, the X-ray crystal structure of the human MLKL pseudokinase reported by Murphy and co-workers (Figure 3D) was revealed as monomeric protein that lacks two of the three catalytic residues [36]. Nevertheless, native MLKL is able to bind ATP, ADP, GTP, and AMP-PNP in vitro in the absence of divalent ions with an affinity for ATP. The addition of divalent ions drastically decreases the affinity for nucleotide substrates [36]. Furthermore, MLKL illustrates the versatility of a pseudokinase domain, which acts as a molecular switch, as a suppressor of the 4HB executioner domain [37], and as a protein interaction domain to recruit downstream effectors, such as the Cdc37-Hsp90 [38].

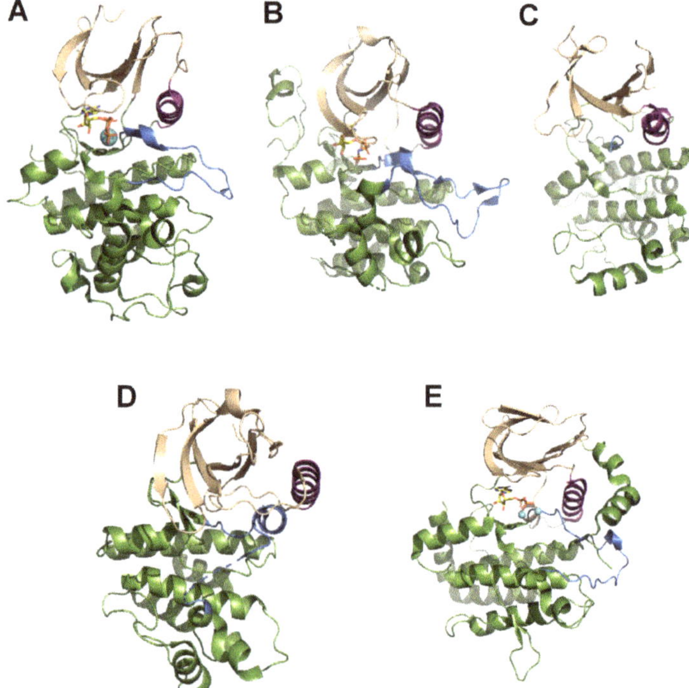

Figure 3. Comparison of the 3D structure of (**A**) Cdk2 kinase (PDB ID 1QMZ); and representatives pseudokinases (**B**) Stradα, (PDB ID 2WTK); (**C**) Trib1 (PDB ID 5CEM); and (**D**) MLKL (PDB ID 4BTF). In the human, BubR1 seems to function as a pseudokinase. However, the recently reported crystal structure of BubR1 from the fruit fly (Panel E, PDB ID 6JKM) revealed a bona fide protein kinase domain. In each structure, highlighted in colour is the N-terminal lobe (salmon); the alpha-helix (purple) and the activation loop (marine). In the Cdk2 (**A**) and BubR1 (**E**) structures, magnesium ions are shown in sphere representation.

In the noncatalytic pseudokinase Stradα, also known as Liver Kinase B1 (LKB1), the pseudo-active site lacks the three catalytic residues (K residue in the VAIK motif, D residue in the HRD motif and a D residue in the DFG motif). Nevertheless, Stradα still binds ATP in an orientation similar to that observed in bona fide protein kinases such as Cdk2 (Figure 3A), including the coordination of the nucleotide by a lysine/tyrosine amino acid residue [34] (Figure 3B). In stark contrast, other pseudokinases such as vaccinia-related kinase 3 (VRK3) appear to have lost the ability to bind ATP [30].

(c) pseudokinases bind cations but not nucleotides. Only the proteins ROP2 (pdb: 2w1z, 3dzo) and PEAK1 (pdb: 6bhc) have been described to bind divalent cations but not nucleotides [39,40]. The crystal structures of both proteins revealed a highly occluded binding site with the DFG and YRDLKPEN motifs clearly visible and the cation binding residues are located in a region of low structural complexity.

(d) pseudokinases bind nucleotides and cations. This group includes the proteins ADCK3 (pdb: 4ped); BSK8 (pdb:4l92, 4l93, 4l94); HER3 (pdb:3kex, 3lmg, 4otw, 4riw, 4rix, 4riy); HSER, ILK (pdb: 3kmu, 3kmw, 3rep, 6mib); IRAK2, KSR1, KSR2 (pdb: 2y4l, 5kkr); PAN3 (pdb: 4bwk, 4bwx, 4bwp, 4cyl, 4cyj, 4czy, 4xr7); RNase L (pdb: 4oav, 4oau, 4o1p, 4o1o); ROP5B (pdb: 3q5z, 3q60, 4lv5); STKLD1, Tyk2jh2 (pdb: 3zon, 4oli, 4wov, 5tkd); and Jak1-2 (pdb: 4l00, 4l01, 4fvp, 4fvq, 4fvr, 5l4n, 5usz, 5uto, and 5ut1-6). Most of these proteins have an intact DFG motif and present kinase activity slightly lower than that of active kinases.

The pseudokinase HER3, a member of the epidermal EGFR family of receptor tyrosine kinases, lacks the canonical catalytic Asp residue although its able to bind tightly to ATP. Ligand-induced heterodimerisation of HER3 with EGFR and HER2 modulates activation of the PI3K/AKT signalling pathway [41,42].

The JAK kinases family (JAK1, JAK2, Jak3 and TYK2) contains a FERM domain, a SH2-like domain and a pseudokinase (JH2) domain adjacent to a C-terminal tyrosine-kinase domain (JH1) (Figure 4) [43]. JAK2 is able to autophosphorylate the residues Ser523 and Tyr570 [44] and residue substitutions in the nucleotide binding site prevent autophosphorylation of both residues.

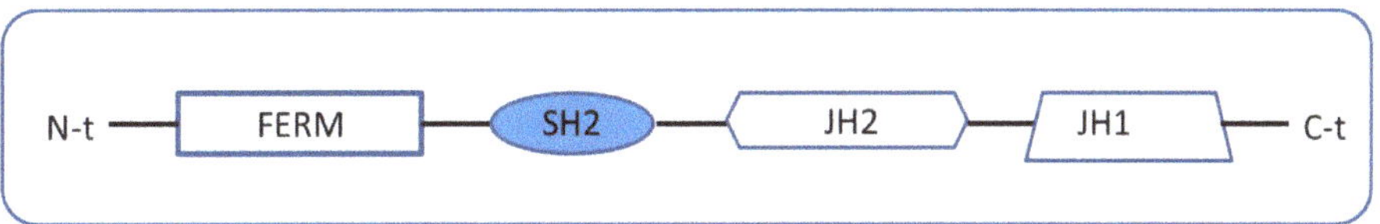

Figure 4. Domain organisation of JAKs pseudokinases.

In addition to JAK2, the JAK family of nonreceptor tyrosine kinases consists of JAK1, JAK3 and TYK2 [45], which share seven regions of sequence termed Janus homology (JH) domains. JH1 consists of a conventional tyrosine kinase domain that becomes activated upon stimulation of type I/II cytokine receptors, and mediates a variety of biological processes including hematopoiesis and immune response regulation [46]. The JH2 domain lacks the catalytic Asp residue of the HRD motif but still appears to regulate Jak2 signalling through ATP binding and/or weak catalytic activity [44]. The tyrosine kinase domain of all JAKs is suppressed on co-expression with the pseudokinase domain, either independently or in tandem. Mutations causing activation of the tyrosine kinase domain occur in haematological malignancies [47], whereas loss-of-function mutations have been identified in immune deficiencies [48]. Despite these advances, the precise molecular mechanism by which the JAK pseudokinase domains suppress the activity of the tyrosine kinase domain remains to be established. In addition to Janus kinases, one pseudokinase that regulates other classes of enzymes is vaccinia-related kinase 3 (VRK3), which is known to allosterically regulate the Erk phosphatase DUSP3/VHR [49].

The pseudokinases KSR1 (RAS1) and KSR2 (RAS2) act as scaffolding proteins that organise the assembly of a Raf–MEK–ERK complex, which drives oncogenesis [50–52]. KSR1 and 2 are also important regulators of central metabolic pathways and the immune system [53,54]. KSR2 disease-associated mutations disrupt KSR2-mediated signalling via the Raf–MEK–ERK pathway and have been implicated in obesity, insulin resistance and impaired cellular fuel oxidation [55]. While IRAK1 and IRAK4 are bona fide protein kinases, IRAK2 and IRAK3 are classified as pseudokinases. IRAK2 is a positive regulator of NF-κB pro-inflammatory signalling by promoting TRAF6 polyubiquitylation [56,57] and mutations within the pseudokinase domain can either enhance or suppress the magnitude of this signalling response. IRAK3 binds to and antagonises other IRAK family members, acting as a switch to initiate suppression of the inflammatory response implicating NF-κB signalling [56,58,59].

3. Mechanism of Action

The catalytic mechanism of bona fide protein kinases is highly conserved across the tree of life and involves the following steps: (a) nucleotide binding to the active site of the protein, (b) the transfer of the γ-phosphate of the nucleotide to a hydroxyl group of the Ser, Thr, or Tyr residues of the protein and (d) the release of ADP from the active site of the phosphorylated protein (Figure 5).

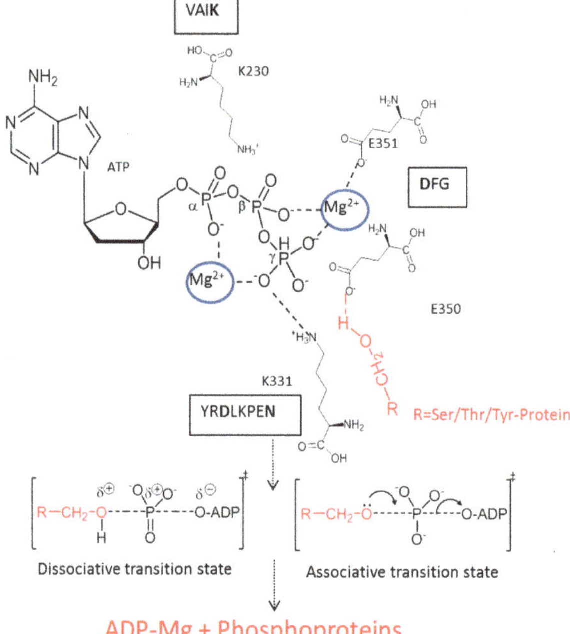

Figure 5. General catalytic mechanism of catalytically active protein kinases. The relative position of the conserved motifs VAIK, YRDLKPEN and DFG is indicated.

In contrast, pseudokinases can employ different modes of action: (a) modulating the activity of kinases by serving as dimerisation partners, thereby inhibiting or accelerating kinase activity perhaps through an allosteric transition, (b) competition for substrates of protein kinases, (c) spatial anchor to trap a substrate into particular subcellular location and (d) as a signalling hub that mediates interactions with components of different signalling pathways, or with multiple components of a linear signalling cascade [11] (Figure 6).

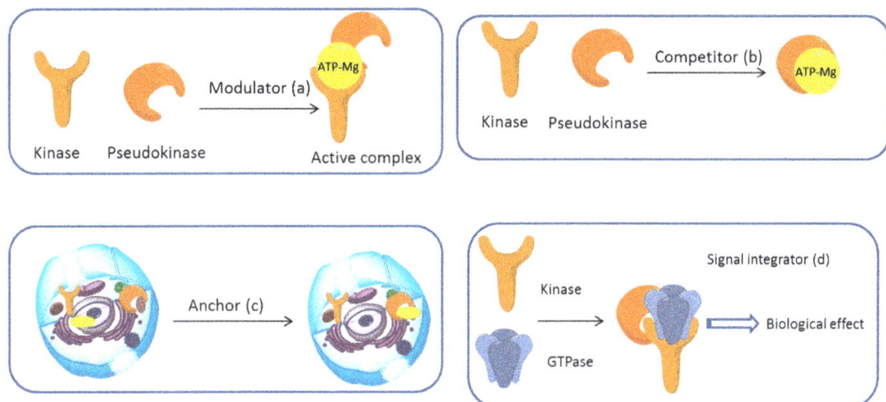

Figure 6. Modes of action of pseudokinases; (**a**) dimerisation (**b**) competitor (**c**) anchor and (**d**) signal integrator.

4. Pseudokinases and Subcellular Localisation

4.1. Transmembrane Pseudokinases

Nearly 25% of the pseudokinases reported to date contain a functional transmembrane domain. Of these, the homodimeric receptor guanylyl cyclases (RGCs) constitute the largest subgroup, in which the pseudokinase domain occupies the cytoplasmic segment located between the transmembrane domain and a functional guanylyl cyclase domain [60]. The pseudokinase domains associate with each other following ligand binding, which in turn activate the guanylyl cyclase domains probably through an allosteric mechanism [61].

4.2. Nuclear Pseudokinases

The pseudokinase nuclear receptor-binding protein 1 (NRBP1) functions as adaptor and contains a putative Src homology 2 and nuclear receptor-binding domains, as well as sequences that drive protein nuclear import and export [62]. Also in the nucleus resides the Tribbles pseudokinase (TRIB) protein family, whose functions have been described in Section 2. At least in mitosis, the pseudokinase Budding uninhibited by benzimidazoles related 1 (BubR1), functions as a pseudokinase that plays important roles in chromosome segregation but also has been reported to play important functions in DNA repair, ciliogenesis and neuron differentiation [63]. BubR1 together with Budding uninhibited by benzimidazoles 3 (Bub3), Mitotic arrest deficient 2 (Mad2) and Cdc20 assemble to form the Mitotic Checkpoint Complex (MCC), an assembly that regulates the E3 ubiquitin ligase activity of the Anaphase Promoting Complex/Cyclosome (APC/C) [64,65]. BubR1 also functions as a mechano sensor between kinetochores and microtubules [66,67]. Five main regions can be identified in the BubR1 polypeptide chain: (i) two units of the KEN box motif located in the N-terminal region, and one putative destruction box (D-box) motif located in the C-terminal region; (ii) a N-terminal fragment that is organised as a triple-tandem arrangement of the tetratricopeptide repeat (TPR) motif that contributes to the kinetochore localisation of BubR1; (iii) an intermediate, non-conserved region of low structural complexity that is required for the binding to Bub3; (iv) a region harbouring a Cdc20 binding site; and (v) a C-terminal region that contains a serine/threonine kinase domain.

In the human, BubR1 has been reported to function as a pseudokinase [68]. However, at least in the fruit fly the BubR1 kinase domain is catalytically active and phosphorylates the kinetochore protein CENP-E [69]. The phosphorylation of this protein by BubR1 is required for spindle microtubules transition from lateral association to end-on capture. The fact that BubR1 seems to act as a pseudokinase in some species and as a catalytically proficient enzyme in others exemplifies the current challenges in the identification and characterisation of pseudokinases. Other important nuclear pseudokinases are the transformation/transcription domain-associated protein (TRRAP), which functions as a scaffold platform for the assembly of the histone acetyltransferase complex [70–74] and Trib1, which induces nuclear retention of COP1, a highly conserved ubiquitin ligase that regulates diverse cellular processes in plants and metazoans [75].

4.3. Cytoplasmic Pseudokinases

Other pseudokinases are involved in intracellular signalling hubs predominantly in the cytoplasm such as Titin (TTN), a protein firstly characterised as a member of muscle fibre sarcomeres. The N-terminal kinase-like domain of TTN interacts with the zinc finger protein Nbr1, and mutations within the pseudokinase domain that interrupt this interaction are linked to hereditary muscle disease [76]. The TTN pseudokinase domain also recruits the E3 ligases Murf1 and Murf2 [77], and is also proposed to act as a mechanosensor, similar to BubR1 [78]. Another cytoplasmic pseudokinase of particular interest is SCY1-like (SCYLs), which has been implicated in the regulation of intracellular trafficking [79].

5. Pseudokinases and Disease

The use of the bioinformatics resource ProKino (Protein Kinase Ontology Browser) [80] allowed us to map 55 of the human pseudokinases to genes from the open targets project revealing the most enriched disease association to be neurodegenerative disease (p-value = 2e − 9). The most strongly associated gene with this was TBCK. Given the tractability of pseudokinases as drug targets, TBCK appears to be a good drug target that, to the best of our knowledge, remains to be actively pursued. Further, bioinformatics analysis of the 55 human pseudokinases mentioned above indicates that the pathway with the most PKs in the open targets resource is that mediated by Interleukins (p-value = 0.000078).

Although many pseudokinases show a concentration of mutations in the C-terminal subdomain such as Trib1 (Figure 7), other kinases have enrichments at positions other than the kinase domain. One dramatic case is WNK Lysine Deficient Protein Kinase 2 (WNK2) with 38 mutations mapped onto position 1838. The WNK family of Ser/Thr pseudokinases comprises the proteins WNK1 to WNK4 [81] and lacks the conserved β3 lysine assumed to be indispensable for nucleotide binding and stabilisation of the active kinase conformation [82]. Despite this, WNK can regulate diverse intracellular substrates in a phosphorylation-dependent fashion in a process that involves a novel mechanism of catalysis. Namely, the terminal residue in the glycine-rich loop (often a Gly in kinases) is conserved as a Lys residue in WNKs and this residue provides the compensatory charge that is required for productive ATP binding [81]. Neuronal WNK isoforms are associated with hereditary neuropathy and glioma [82]. WNKs are also associated with the control of blood pressure through regulation of SPAK [83,84]. In brief, examples of pseudokinase mutations that lead to impairment of function can be mapped onto the kinase and other adjacent domains and across higher organisms. In the human, such mutations occur in genes encoding for proteins with important roles in metabolic and cell signalling pathways and are associated with metabolic, neurological, autoimmune, and cell proliferation disorders.

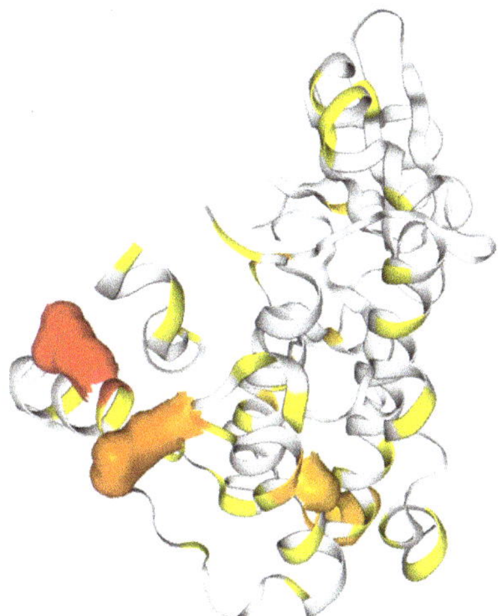

Figure 7. Trib1 is an example of a pseudokinase that contains subregions recurrently mutated. Analysis based on the cosmic 3D Mutation bioinformatics resource.

6. Pseudokinases Protein–Protein Interaction Analysis

Interaction analyses indicate that pseudokinases have great potential for drug innovations. Taking the full network from IntAct EBI in the spoke model, it can be seen that pseudokinases form one main connected component. Classifying genes strongly implicated in cancer and other diseases facilitated the identification of a large number of links between pseudokinases and their interacting partners (show as orange links in Figure 8). In the network, colours of nodes represent pseudokinases from the same family except for the red nodes, which indicates proteins that link two or more different pseudokinases. Colouring the nodes in this way reveals that some pseudokinases from the same family share common interaction partners. This interactor sharing is particularly the case for the TK family members where both TYK2 and JAK1 and ERBB3 and PTK7 share a large number of interactors, suggesting co-regulation between these pseudokinases. It is possible to zoom in on the network and to select only those interactions where both genes are disease-associated and either already confirmed as druggable or with strong potential for druggability as designated by the HPA (Human Protein Atlas, 31 July 2019). Gold edges are indicative of potential for druggable interaction therapeutic intervention because one or both of the partners is a known disease gene.

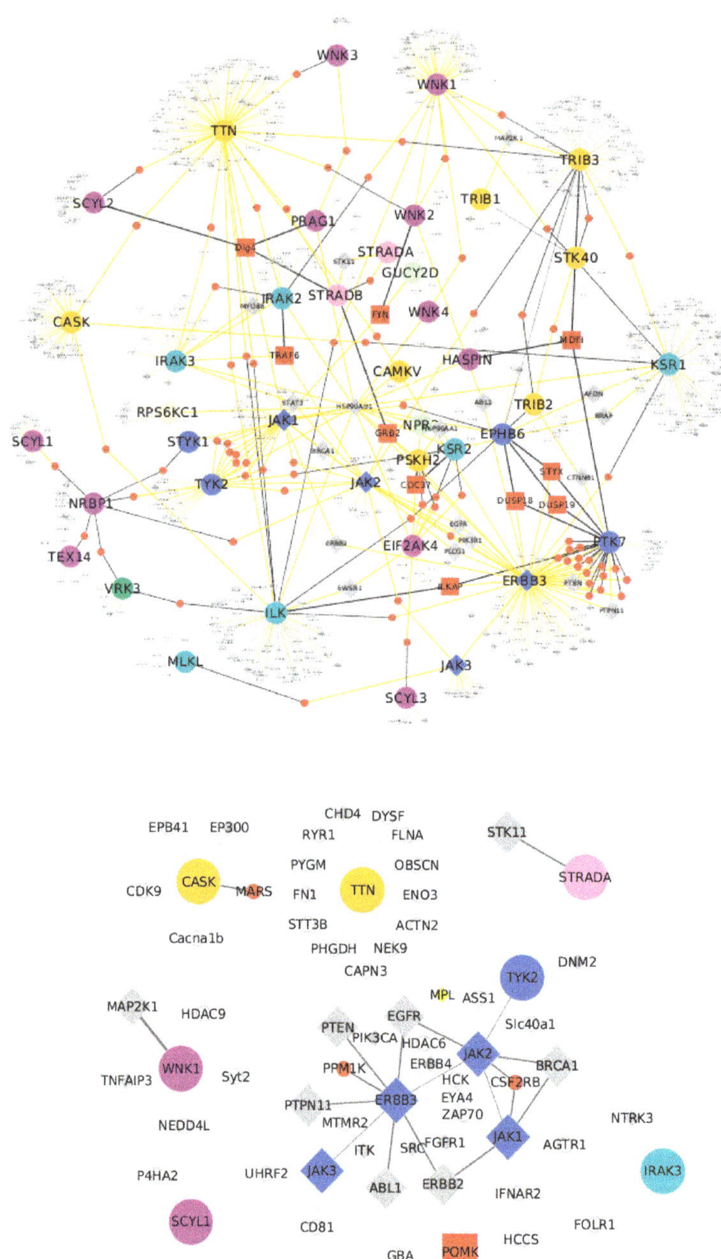

Figure 8. Above, the human pseudokinase interactome. The colours represent family classes. Protein kinase linking nodes are shown in red whereas diamonds indicate HPA genes that are strongly implicated in cancer. Gold edges indicate a potentially therapeutic interaction to target where one of the genes is a cancer or disease gene. Below, a highly druggable sub-network where all nodes are disease-associated genes and have an FDA (Food and Drug Administration) drug or a high potential for druggability as defined by the Human Protein Atlas (HPA) datasets.

To understand better which regions of the pseudokinases were responsible for physical interactions, we extracted interaction features from Intact restricted to feature lengths ≤20. Results in 9 of the 17 pseudokinases showed interaction features that overlap with the kinase domain in its interactors, indicating that interactions through the kinase domain are often important. Conversely, non-PK regions look equally important because only 4 of the 17 kinases always interacted through the pseudokinase domain.

Pseudokinases with residual or total lack of catalytic activity can still carry out important roles in cell signalling involving pseudo-active sites, allosteric transitions driven by nucleotide binding, and/or direct modulation of binding surfaces topology. Because some promiscuous inhibitors that target the ATP-binding site of bona fide protein kinases can bind the equivalent site in pseudokinases [36], the features of the kinase inhibitors can inform the development of new small size drugs to target pseudokinases. The computational protein–protein interaction analyses described in this contribution together with recent approaches in drug discovery, such as the targeting of specific protein–protein interfaces with small size compounds [85–87], further support the view that pseudokinases offer great potential for drug innovations.

7. Pseudokinases as Drug Targets

Because the vast repertoires of functions exerted by pseudokinases represent a new window to develop innovative strategies for therapeutic intervention to treat human diseases that constitute important yet unmet clinical needs [13–15,88–90], we mapped pseudokinases to opentargets to visualise their ability to be targeted by small size drugs or antibodies. We carried out an analysis of their druggability in open targets, which showed that a number of pseudokinases can be potential targets of different drug molecule types (e.g., antibody/small molecule) with the Janus kinases representing the most actively pursued targets (Table 1). Our bioinformatics analyses (Figure 9) strongly suggest that pseudokinases have great potential for drug development and revealed that the top diseases targeted are rheumatoid arthritis (five drugs) followed by neoplasm (four drugs). These findings are in good agreement with recent reports by others, which showed that the unique conformations of Tribbles pseudokinases make these proteins good drug targets to treat human acute megakaryocytic leukemia caused by a gain-of-function mutation [91]; that inhibitors of WNK such as WNK463 have potential use in the clinic as antihypertensive drugs [92]; and that a new generation of TYK2 pseudokinase ligands currently in clinical trials may be effective for the treatment of autoimmune diseases [93,94]. An exciting new development in drug discovery is the use of artificial intelligence (AI) to speed up the process and reduce costs by facilitating the rapid identification of drug-like compounds. A recent example of the promise of AI to accelerate the development of new drugs is the identification of potent inhibitors of the Discoidin domain receptor 1 (DDR1) tyrosine kinase using machine learning techniques [95]. It can be anticipated that a similar approach will be used to identify pseudokinase inhibitors, including drugs that interfere with specific protein–protein interactions.

Table 1. Existing status for pseudokinases in clinical trials as reported in the open targets database.

Target	Max Phase	Molecula Type	Drugs
JAK2	Phase III	Small molecule	5
ERBB3	Phase II	Antibody	4
JAK3	Phase II	Small molecule	4
JAK1	Phase II	Small molecule	3
JAK2	Phase I	Small molecule	3
JAK1	Phase III	Small molecule	3
JAK2	Phase II	Small molecule	3
JAK2	Phase IV	Small molecule	2

Table 1. *Cont.*

Target	Max Phase	Molecula Type	Drugs
ERBB3	Phase IV	Small molecule	2
GUCY2C	Phase IV	protein	2
ERBB3	Phase I	Antibody	2
EPHB6	Phase IV	Small molecule	1
JAK1	Phase IV	Small molecule	1
ERBB3	Phase III	Small molecule	1
ERBB3	Phase III	Antibody	1
ERBB3	Phase II	Small molecule	1
NPR1	Phase IV	protein	1

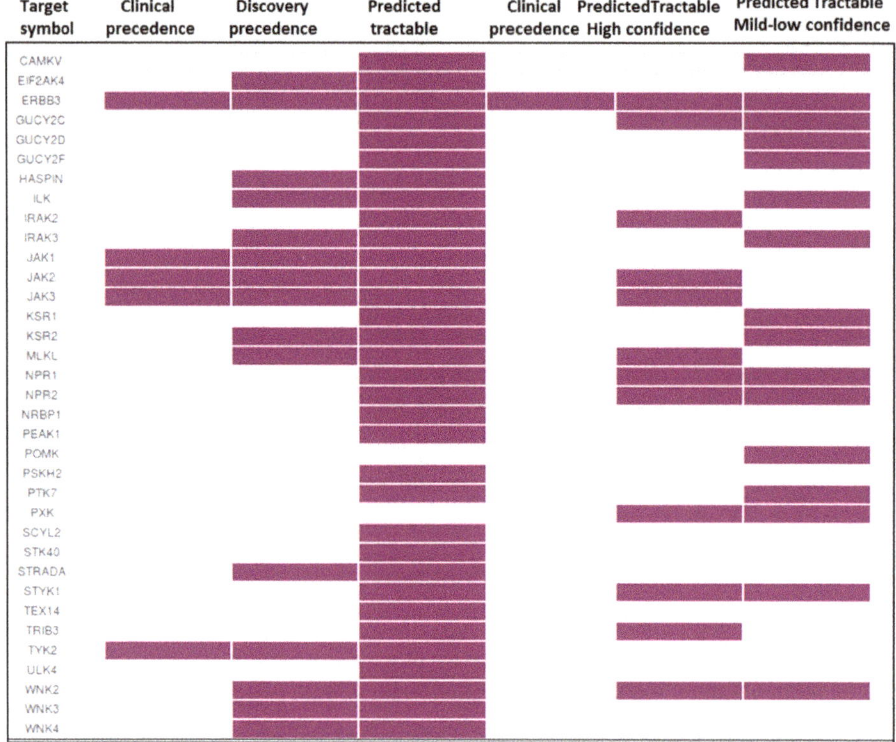

Figure 9. Pseudokinases mapped to opentargets tractability for both small size drug compounds and antibodies.

8. Conclusions

Pseudokinases are a subset of the protein kinase superfamily that present inactivating mutations in critical catalytic motifs but signal primarily through noncatalytic mechanisms. Systematic analysis of the human genome has revealed that circa one-tenth of all protein kinases correspond to pseudokinases. These proteins play central roles in the cell and the loss of their regulation can lead to a variety of diseases. Our succinct analysis of the pseudokinase interactome revealed that TYK2, JAK1, ERBB3, and PTK7 pseudokinases share a large number of interactors, suggesting that they are subjected to co-regulation. Several clinically approved kinase inhibitors have been shown to influence the noncatalytic functions of active kinases, providing expectation that similar properties in pseudokinases could be pharmacologically regulated. Recent advances in drug discovery assisted by the use of artificial intelligence approaches may pave the way for the rapid identification of potent pseudokinase inhibitors. Taken these advances together, it can be anticipated that in the coming years pseudokinases are likely to become mainstream drug targets for the treatment of diverse malignancies.

Author Contributions: Conceptualization: V.M.B.-G., J.L. and A.B.; methodology: V.M.B.-G., J.L. and A.B.; software: J.L.; validation: V.M.B.-G., J.L. and A.B.; formal analysis: V.M.B.-G., J.L. and A.B.; investigation: A.T., V.M.B.-G., J.L., A.G.S and A.B.; resources: V.M.B.-G., J.L. and A.B.; writing—original draft preparation: A.T., V.M.B.-G., J.L., A.G.S and A.B.; writing—review and editing: V.M.B.-G., J.L. and A.B.; supervision: V.M.B.-G., J.L. and A.B.; project administration: V.M.B.-G., J.L. and A.B.; funding acquisition: V.M.B.-G. and A.B.

Funding: This research received no external funding.

Conflicts of Interest: The authors declare no conflict of interest.

References

1. Manning, G.; Whyte, D.B.; Martinez, R.; Hunter, T.; Sudarsanam, S. The protein kinase complement of the human genome. *Science* **2002**, *298*, 1912–1934. [CrossRef] [PubMed]
2. Caenepeel, S.; Charydczak, G.; Sudarsanam, S.; Hunter, T.; Manning, G. The mouse kinome: Discovery and comparative genomics of all mouse protein kinases. *Proc. Natl. Acad. Sci. USA* **2004**, *101*, 11707–11712. [CrossRef] [PubMed]
3. Plowman, G.D.; Sudarsanam, S.; Bingham, J.; Whyte, D.; Hunter, T. The protein kinases of *Caenorhabditis elegans*: A model for signal transduction in multicellular organisms. *Proc. Natl. Acad. Sci. USA* **1999**, *96*, 13603–13610. [CrossRef] [PubMed]
4. Giamas, G.; Man, Y.L.; Hirner, H.; Bischof, J.; Kramer, K.; Khan, K.; Ahmed, S.S.; Stebbing, J.; Knippschild, U. Kinases as targets in the treatment of solid tumors. *Cell. Signal.* **2010**, *22*, 984–1002. [CrossRef] [PubMed]
5. Kwon, A.; Scott, S.; Taujale, R.; Yeung, W.; Kochut, K.J.; Eyers, P.A.; Kannan, N. Tracing the origin and evolution of pseudokinases across the tree of life. *Sci. Signal.* **2019**, *12*, eaav3810. [CrossRef] [PubMed]
6. Manning, G.; Reiner, D.S.; Lauwaet, T.; Dacre, M.; Smith, A.; Zhai, Y.; Svard, S.; Gillin, F.D. The minimal kinome of Giardia lamblia illuminates early kinase evolution and unique parasite biology. *Genome Biol.* **2011**, *12*, R66. [CrossRef] [PubMed]
7. Ribeiro, A.J.M.; Das, S.; Dawson, N.; Zaru, R.; Orchard, S.; Thornton, J.M.; Orengo, C.; Zeqiraj, E.; Murphy, J.M.; Eyers, P.A. Emerging concepts in pseudoenzyme classification, evolution, and signaling. *Sci. Signal.* **2019**, *12*, eaat9797. [CrossRef] [PubMed]
8. Sharir-Ivry, A.; Xia, Y. Using Pseudoenzymes to Probe Evolutionary Design Principles of Enzymes. *Evol. Bioinform.* **2019**, *15*, 1176934319855937. [CrossRef]
9. Jeffery, C.J. The demise of catalysis, but new functions arise: Pseudoenzymes as the phoenixes of the protein world. *Biochem. Soc. Trans.* **2019**, *47*, 371–379. [CrossRef]
10. Bailey, F.P.; Byrne, D.P.; McSkimming, D.; Kannan, N.; Eyers, P.A. Going for broke: Targeting the human cancer pseudokinome. *Biochem. J.* **2015**, *465*, 195–211. [CrossRef]
11. Reiterer, V.; Eyers, P.A.; Farhan, H. Day of the dead: Pseudokinases and pseudophosphatases in physiology and disease. *Trends Cell Biol.* **2014**, *24*, 489–505. [CrossRef] [PubMed]
12. Zhang, H.; Photiou, A.; Grothey, A.; Stebbing, J.; Giamas, G. The role of pseudokinases in cancer. *Cell. Signal.* **2012**, *24*, 1173–1184. [CrossRef] [PubMed]
13. Boudeau, J.; Miranda-Saavedra, D.; Barton, G.J.; Alessi, D.R. Emerging roles of pseudokinases. *Trends Cell Biol.* **2006**, *16*, 443–452. [CrossRef] [PubMed]
14. Eyers, P.; Murphy, J.M. Dawn of the dead: Protein pseudokinases signal new adventures in cell biology. *Biochem. Soc. Trans.* **2013**, *41*, 969–974. [CrossRef] [PubMed]
15. Kung, J.E.; Jura, N. Prospects for pharmacological targeting of pseudokinases. *Drug Dev. Nat. Rev. Drug Discov.* **2019**, *18*, 501–526. [CrossRef] [PubMed]
16. Fabbro, D.; Cowan-Jacob, S.W.; Moebitz, H. Ten things you should know about protein kinases: IUPHAR Review 14. *Br. J. Pharmacol.* **2015**, *172*, 2675–2700. [CrossRef] [PubMed]
17. Salazar, M.; Lorente, M.; Orea-Soufi, A.; Dávila, D.; Erazo, T.; Lizcano, J.; Carracedo, A.; Kiss-Toth, E.; Velasco, G. Oncosuppressive functions of tribbles pseudokinase 3. *Biochem. Soc. Trans.* **2015**, *43*, 1122–1126. [CrossRef]

18. Wu, P.; Nielsen, T.E.; Clausen, M.H. FDA-approved small-molecule kinase inhibitors. *Trends Pharmacol. Sci.* **2015**, *36*, 422–439. [CrossRef]

19. Knight, Z.A.; Shokat, K.M. Features of Selective Kinase Inhibitors. *Chem. Biol.* **2005**, *12*, 621–637. [CrossRef]

20. Davis, M.I.; Hunt, J.P.; Herrgard, S.; Ciceri, P.; Wodicka, L.M.; Pallares, G.; Hocker, M.; Treiber, D.K.; Zarrinkar, P.P. Comprehensive analysis of kinase inhibitor selectivity. *Nat. Biotechnol.* **2011**, *29*, 1046–1051. [CrossRef]

21. Zhang, X.; Gureasko, J.; Shen, K.; Cole, P.A.; Kuriyan, J. An allosteric mechanism for activation of the kinase domain of epidermal growth factor receptor. *Cell* **2006**, *125*, 1137–1149. [CrossRef] [PubMed]

22. Littlefield, P.; Liu, L.; Mysore, V.; Shan, Y.; Shaw, D.E.; Jura, N. Structural analysis of the EGFR/HER3 heterodimer reveals the molecular basis for activating HER3 mutations. *Sci. Signal.* **2014**, *7*, ra114. [CrossRef] [PubMed]

23. Rajakulendran, T.; Sicheri, F. Allosteric protein kinase regulation by pseudokinases: Insights from STRAD. *Sci. Signal.* **2010**, *3*, pe8. [CrossRef] [PubMed]

24. Vulpetti, A.; Bosotti, R. Sequence and structural analysis of kinase ATP pocket residues. *Farmaco* **2004**, *59*, 759–765. [CrossRef] [PubMed]

25. Huse, M.; Kuriyan, J. The conformational plasticity of protein kinases. *Cell* **2002**, *109*, 275–282. [CrossRef]

26. Taylor, S.S.; Kornev, A.P. Protein kinases: Evolution of dynamic regulatory proteins. *Trends Biochem. Sci.* **2011**, *36*, 65–77. [CrossRef] [PubMed]

27. Bose, R.; Holbert, M.A.; Pickin, K.A.; Cole, P.A. Protein tyrosine kinase-substrate interactions. *Curr. Opin. Struct. Biol.* **2006**, *16*, 668–675. [CrossRef]

28. Murphy, J.M.; Zhang, Q.; Young, S.N.; Reese, M.L.; Bailey, F.P.; Eyers, P.A.; Ungureanu, D.; Hammaren, H.; Silvennoinen, O.; Varghese, L.N.; et al. A robust methodology to subclassify pseudokinases based on their nucleotide-binding properties. *Biochem. J.* **2013**, *457*, 323–334. [CrossRef]

29. Jacobsen, A.V.; Murphy, J.M. The secret life of kinases: Insights into non-catalytic signalling functions from pseudokinases. *Biochem. Soc. Trans.* **2017**, *15*, 665–681. [CrossRef]

30. Scheeff, E.D.; Eswaran, J.; Bunkoczi, G.; Knapp, S.; Manning, G. Structure of the pseudokinase VRK3 reveals a degraded catalytic site, a highly conserved kinase fold, and a putative regulatory binding site. *Structure* **2009**, *17*, 128–138. [CrossRef]

31. Bailey, F.P.; Byrne, D.P.; Oruganty, K.; Eyers, C.E.; Novotny, C.J.; Shokat, K.M.; Kannan, N.; Eyers, P.A. The Tribbles 2 (TRB2) pseudokinase binds to ATP and autophosphorylates in a metal independent manner. *Biochem. J.* **2015**, *467*, 47–62. [CrossRef] [PubMed]

32. Foulkes, D.M.; Byrne, D.P.; Bailey, F.P.; Eyers, P.A. Tribbles pseudokinases: Novel targets for chemical biology and drug discovery? *Biochem. Soc. Trans.* **2015**, *43*, 1095–1103. [CrossRef] [PubMed]

33. Eyers, P.A.; Keeshan, K.; Kannan, N. Tribbles in the 21st Century: The evolving roles of tribbles pseudokinases in biology and disease. *Trends Cell Biol.* **2017**, *27*, 284–298. [CrossRef] [PubMed]

34. Zeqiraj, E.; Filippi, B.M.; Goldie, S.; Navratilova, I.; Boudeau, J.; Deak, M.; Alessi, D.R.; van Aalten, D.M.F. ATP and MO25α regulate the conformational state of the STRADα pseudokinase and activation of the LKB1 tumour suppressor. *PLoS Biol.* **2009**, *7*, 1000126. [CrossRef] [PubMed]

35. Cui, J.; Zhu, Q.; Zhang, H.; Cianfrocco, M.A.; Leschziner, A.E.; Dixon, J.E.; Xiao, J. Structure of Fam20A reveals a pseudokinase featuring unique disulfide pattern and inverted ATP-binding. *eLife* **2017**, *6*, 1–16. [CrossRef] [PubMed]

36. Murphy, J.M.; Lucet, I.S.; Hildebrand, J.M.; Tanzer, M.C.; Young, S.N.; Sharma, P.; Lessene, G.; Warren, S.A.; Babon, J.J.; Silke, J.; et al. Insights into the evolution of divergent nucleotide-binding mechanisms among pseudokinases revealed by crystal structures of human and mouse MLKL. *Biochem. J.* **2014**, *457*, 369–377. [CrossRef]

37. Petrie, E.J.; Sandow, J.J.; Jacobsen, A.V.; Smith, B.J.; Griffin, M.D.W.; Lucet, I.S.; Dai, W.; Young, S.N.; Tanzer, M.C.; Wardak, A.; et al. Conformational switching of the pseudokinase domain promotes human MLKL tetramerization and cell death by necroptosis. *Nat. Commun.* **2018**, *9*, 2422. [CrossRef]

38. Jacobsen, A.V.; Lowes, K.N.; Tanzer, M.Z.; Lucet, I.S.; Hildebrand, J.M.; Petrie, E.J.; van Delft, M.F.; Liu, Z.; Conos, S.A.; Zhang, J.-G.; et al. HSP90 activity is required for MLKL oligomerisation and membrane translocation and the induction of necroptotic cell death. *Cell Death Dis.* **2016**, *7*, e2051. [CrossRef]

39. Ha, B.H.; Boggon, T.J. The crystal structure of pseudokinase PEAK1 (Sugen Kinase 269) reveals an unusual catalytic cleft and a novel mode of kinase fold dimerization. *J. Biol. Chem.* **2017**, *293*, 1642–1650. [CrossRef]

40. Labesse, G.; Gelin, M.; Bessin, Y.; Lebrun, M.; Papoin, J.; Cerdan, R.; Arold, S.T.; Dubremetz, J.F. ROP2 from *Toxoplasma gondii*: A virulence factor with a protein- kinase foldand no enzymatic activity. *Structure* **2009**, *17*, 139–146. [CrossRef]

41. Yarden, Y.; Sliwkowski, M.X. Untangling the ErbB signaling network. *Nat. Rev. Mol. Cell. Biol.* **2001**, *2*, 127–137. [CrossRef] [PubMed]

42. Zhang, N.; Chang, Y.; Rios, A.; An, Z. HER3/ErbB3, an emerging cancer therapeutic target. *Acta Biochim. Biophys. Sin.* **2016**, *48*, 39–48. [CrossRef] [PubMed]

43. Toms, A.V.; Deshpande, A.; McNally, R.; Jeong, Y.; Rogers, J.M.; Kim, C.U.; Gruner, S.M.; Ficarro, S.B.; Marto, J.A.; Sattler, M.; et al. Structure of a pseudokinase domain switch that controls oncogenic activation of Jak kinases. *Nat. Struct. Mol. Biol.* **2014**, *20*, 1221–1223. [CrossRef] [PubMed]

44. Ungureanu, D.; Wu, J.; Pekkala, T.; Niranjan, Y.; Young, C.; Jensen, O.N.; Xu, C.F.; Neubert, T.A.; Skoda, R.C.; Hubbard, S.R.; et al. The pseudokinase domain of JAK2 is a dual-specificity protein kinase that negatively regulates cytokine signaling. *Nat. Struct. Mol. Biol.* **2011**, *14*, 971–976. [CrossRef] [PubMed]

45. Laurence, A.; Pesu, M.; Silvennoinen, O.; O'shea, J. JAK kinases in health and disease: An update. *Open Rheumatol. J.* **2012**, *6*, 232–244. [CrossRef]

46. Haan, C.; Behrmann, I.; Haan, S. Perspectives for the use of structural information and chemical genetics to develop inhibitors of Janus kinases. *J. Cell. Mol. Med.* **2010**, *14*, 504–527. [CrossRef] [PubMed]

47. Hammarén, H.M.; Ungureanu, D.; Grisouard, J.; Skoda, R.C.; Hubbard, S.R.; Silvennoinena, O. ATP binding to the pseudokinase domain of JAK2 is critical for pathogenic activation. *Proc. Natl. Acad. Sci. USA* **2015**, *112*, 4642–4647. [CrossRef]

48. Notarangelo, L.D.; Mella, P.; Jones, A.; de Saint Basile, G.; Savoldi, G.; Cranston, T.; Vihinen, M.; Schumacher, R.F. Mutations in severe combined immune deficiency (SCID) due to JAK3 deficiency. *Hum. Mutat.* **2001**, *18*, 255–263. [CrossRef]

49. Amand, M.; Erpicum, C.; Bajou, K.; Cerignoli, F.; Blacher, S.; Martin, M.; Dequiedt, F.; Drion, P.; Singh, P.; Zurashvili, T.; et al. DUSP3/VHR is a pro-angiogenic atypical dual-specificity phosphatase. *Mol. Cancer* **2014**, *13*, 108–126. [CrossRef]

50. Yoon, S.; Seger, R. The extracellular signal-regulated kinase: Multiple substrates regulate diverse cellular functions. *Growth Factors* **2006**, *24*, 21–44. [CrossRef]

51. Rauch, J.; Volinsky, N.; Romano, D.; Kolch, W. The secret life of kinases: Functions beyond catalysis. *Cell Commun. Signal.* **2011**, *9*, 23. [CrossRef] [PubMed]

52. Kolch, W. Coordinating ERK/MAPK signalling through scaffolds and inhibitors. *Nat. Rev. Mol. Cell. Biol.* **2005**, *6*, 827–837. [CrossRef] [PubMed]

53. Costanzo-Garvey, D.L.; Pfluger, P.T.; Dougherty, M.K.; Stock, J.L.; Boehm, M.; Chaika, O.; Fernandez, M.R.; Fisher, K.; Kortum, R.L.; Hong, E.G.; et al. KSR2 is an essential regulator of AMP kinase, energy expenditure, and insulin sensitivity. *Cell Metab.* **2009**, *10*, 366–378. [CrossRef] [PubMed]

54. Revelli, J.P.; Smith, D.; Allen, J.; Jeter-Jones, S.; Shadoan, M.K.; Desai, U.; Schneider, M.; van Sligtenhorst, I.; Kirkpatrick, L.; Platt, K.A.; et al. Profound obesity secondary to hyperphagia in mice lacking kinase suppressor of ras 2. *Obesity* **2011**, *19*, 1010–1018. [CrossRef]

55. Pearce, L.R.; Atanassova, N.; Banton, M.C.; Bottomley, B.; van der Klaauw, A.A.; Revelli, J.P.; Hendricks, A.; Keogh, J.M.; Henning, E.; Doree, D.; et al. KSR2 mutations are associated with obesity, insulin resistance, and impaired cellular fuel oxidation. *Cell* **2013**, *7*, 765–777. [CrossRef] [PubMed]

56. Rhyasen, G.W.; Starczynowski, D.T. IRAK signalling in cancer. *Br. J. Cancer* **2015**, *112*, 232–237. [CrossRef] [PubMed]

57. Lin, S.C.; Lo, Y.C.; Wu, H. Helical assembly in the Myd88-IRAK4-IRAK2 complex in TLR/IL-1R signalling. *Nature* **2010**, *465*, 885–890. [CrossRef] [PubMed]

58. Ge, Z.W.; Wang, B.C.; Hu, J.L.; Sun, J.J.; Wang, S.; Chen, X.J.; Meng, S.P.; Liu, L.; Cheng, Z.Y. IRAK3 gene silencing prevents cardiac rupture and ventricular remodeling through negative regulation of the NF-κB signaling pathway in a mouse model of acute myocardial infarction. *J. Cell Physiol.* **2019**, *234*, 11722–11733. [CrossRef] [PubMed]

59. Kobayashi, K.; Hernandez, L.D.; Galan, J.E.; Janeway, C.A., Jr.; Medzhitov, R.; Flavell, R.A. IRAK-M is a negative regulator of toll-like receptor signaling. *Cell* **2002**, *110*, 191–202. [CrossRef]

60. Kuhn, M. Molecular physiology of membrane guanylyl cyclase receptors. *Physiol. Rev.* **2016**, *96*, 751–804. [CrossRef]

61. Biswas, K.H.; Shenoy, A.R.; Dutta, A.; Visweswariah, S.S. The evolution of guanylyl cyclases as multidomain proteins: Conserved features of kinase-cyclase domain fusions. *J. Mol. Evol.* **2009**, *68*, 587–602. [CrossRef] [PubMed]

62. Kerr, J.S.; Wilson, C.H. Nuclear receptor-binding protein 1: A novel tumour suppressor and pseudokinase. *Biochem. Soc. Trans.* **2013**, *41*, 1055–1060. [CrossRef] [PubMed]

63. Bolanos-Garcia, V.M.; Blundell, T.L. BUB1 and BUBR1: Multifaceted kinases of the cell cycle. *Trends Biochem. Sci.* **2001**, *36*, 141–150. [CrossRef] [PubMed]

64. Hein, J.B.; Nilsson, J. Stable MCC binding to the APC/C is required for a functional spindle assembly checkpoint. *EMBO Rep.* **2014**, *15*, 264–272. [CrossRef] [PubMed]

65. Alfieri, C.; Chang, L.; Zhang, Z.; Yang, J.; Maslen, S.; Skehel, M.; Barford, D. Molecular basis of APC/C regulation by the spindle assembly checkpoint. *Nature* **2016**, *536*, 431–436. [CrossRef] [PubMed]

66. Huang, H.; Hittle, J.; Zappacosta, F.; Annan, R.S.; Hershko, A.; Yen, T.J. Phosphorylation sites in BubR1 that regulate kinetochore attachment, tension, and mitotic exit. *J. Cell Biol.* **2008**, *183*, 667–680. [CrossRef]

67. Chan, G.K.T.; Jablonski, S.A.; Sudakin, V.; Hittle, J.C.; Yen, T.J. Human BUBR1 is a mitotic checkpoint kinase that monitors CENP-E functions at kinetochores and binds the cyclosome/APC. *J. Cell Biol.* **1999**, *146*, 941–954. [CrossRef]

68. Suijkerbuijk, S.J.; van Dam, T.J.; Karagöz, G.E.; von Castelmur, E.; Hubner, N.C.; Duarte, A.M.; Vleugel, M.; Perrakis, A.; Rüdiger, S.G.; Snel, B.; et al. The vertebrate mitotic checkpoint protein BUBR1 is an unusual pseudokinase. *Dev. Cell.* **2012**, *22*, 1321–1329. [CrossRef]

69. Huang, Y.; Lin, L.; Liu, X.; Ye, S.; Yao, P.Y.; Wang, W.; Yang, F.; Gao, X.; Li, J.; Zhang, Y.; et al. BubR1 phosphorylates CENP-E as a switch enabling the transition from lateral association to end-on capture of spindle microtubules. *Cell Res.* **2019**, *29*, 562–578. [CrossRef]

70. Li, H.; Cuenin, C.; Murr, R.; Wang, Z.-Q.; Herceg, Z. HAT cofactor Trrap regulates the mitotic checkpoint by modulation of Mad1 and Mad2 expression. *EMBO J.* **2004**, *23*, 4824–4834. [CrossRef]

71. McMahon, S.B.; Wood, M.A.; Cole, M.D. The essential cofactor TRRAP recruits the histone acetyltransferase hGCN5 to c-Myc. *Mol. Cell. Biol.* **2000**, *20*, 556–562. [CrossRef] [PubMed]

72. Murr, R.; Loizou, J.I.; Yang, Y.-G.; Cuenin, C.; Li, H.; Wang, Z.-Q.; Herceq, Z. Histone acetylation by Trrap-Tip60 modulates loading of repair proteins and repair of DNA double-strand breaks. *Nat. Cell Biol.* **2006**, *8*, 91–99. [CrossRef] [PubMed]

73. Ichim, G.; Mola, M.; Finkbeiner, M.; Cros, M.-P.; Herceg, Z.; Hernandez-Vargas, H. The histone acetyltransferase component TRRAP is targeted for destruction during the cell cycle. *Oncogene* **2014**, *33*, 181–192. [CrossRef] [PubMed]

74. Tapias, A.; Zhou, Z.-W.; Shi, Y.; Chong, Z.; Wang, P.; Groth, M.; Platzer, M.; Huttner, W.; Herceg, Z.; Yang, Y.G.; et al. Trrap-dependent histone acetylation specifically regulates cell-cycle gene transcription to control neural progenitor fate decisions. *Cell Stem Cell* **2014**, *14*, 632–643. [CrossRef] [PubMed]

75. Kung, J.E.; Jura, N. The pseudokinase TRIB1 toggles an intramolecular switch to regulate COP1 nuclear export. *EMBO J.* **2019**, *38*, e99708. [CrossRef] [PubMed]

76. Lange, S.; Xiang, F.; Yakovenko, A.; Vihola, A.; Hackman, P.; Rostkova, E. The kinase domain of Titin controls muscle gene expression and protein turnover. *Science* **2005**, *308*, 1599–1603. [CrossRef]

77. Bogomolovas, J.; Gasch, A.; Simkovic, F.; Rigden, D.J.; Labeit, S.; Mayans, O. Titin kinase is an inactive pseudokinase scaffold that supports MuRF1 recruitment to the sarcomeric M-line. *Open Biol.* **2014**, *4*, 140041. [CrossRef]

78. Puchner, E.M.; Alexandrovich, A.; Kho, A.L.; Hensen, U.; Schafer, L.V.; Brandmeier, B. Mechanoenzymatics of titin kinase. *Proc. Natl. Acad. Sci. USA* **2008**, *105*, 13385–13390. [CrossRef]

79. Pelletier, S. SCYL pseudokinases in neuronal function and survival. *Neural Regen. Res.* **2016**, *11*, 42–44. [CrossRef]

80. A Partnership to Transform Drug Discovery through the Systematic Identification and Prioritisation of Targets. Available online: www.opentargets.org/ (accessed on 22 August 2019).

81. Tang, B.L. (WNK)ing at death: With-no-lysine (Wnk) kinases in neuropathies and neuronal survival. *Brain Res. Bull.* **2016**, *125*, 92–98. [CrossRef]

82. Xu, B.-E.; English, J.M.; Wilsbacher, J.L.; Stippec, S.; Goldsmith, E.J.; Cobb, M.H. WNK1, a novel mammalian serine/threonine protein kinase lacking the catalytic lysine in subdomain II. *J. Biol. Chem.* **2000**, *275*, 16795–16801. [CrossRef] [PubMed]

83. Wu, A.; Wolley, M.; Stowasser, M. The interplay of renal potassium and sodium handling in blood pressure regulation: Critical role of the WNK-SPAK-NCC pathway. *J. Hum. Hypertens.* **2019**, *33*, 508–523. [CrossRef] [PubMed]

84. Shekarabi, M.; Zhang, J.; Khanna, A.R.; Ellison, D.H.; Delpire, E.; Kahle, K.T. WNK Kinase Signaling in Ion Homeostasis and Human Disease. *Cell Metab.* **2017**, *25*, 285–299. [CrossRef] [PubMed]

85. Pommier, Y.; Kiselev, E.; Marchand, C. Interfacial inhibitors. *Bioorg. Med. Chem. Lett.* **2015**, *25*, 3961–3965. [CrossRef] [PubMed]

86. Jin, L.; Wang, W.; Fang, G. Targeting protein-protein interaction by small molecules. *Annu. Rev. Pharmacol. Toxicol.* **2014**, *54*, 435–456. [CrossRef] [PubMed]

87. Jubb, H.; Blundell, T.L.; Ascher, D.B. Flexibility and small pockets at protein-protein interfaces: New insights into druggability. *Prog. Biophys. Mol. Biol.* **2015**, *119*, 2–9. [CrossRef] [PubMed]

88. Karvonen, H.; Perttilä, R.; Niininen, W.; Barker, H.; Ungureanu, D. Targeting Wnt signaling pseudokinases in hematological cancers. *Eur. J. Haematol.* **2018**, *101*, 457–465. [CrossRef] [PubMed]

89. Byrne, D.P.; Foulkes, D.M.; Eyers, P.A. Pseudokinases: Update on their functions and evaluation as new drug targets. *Future Med. Chem.* **2017**, *9*, 245–265. [CrossRef]

90. Cowan-Jacob, S.W.; Jahnke, W.; Knapp, S. Novel approaches for targeting kinases: Allosteric inhibition, allosteric activation and pseudokinases. *Future Med. Chem.* **2014**, *6*, 541–561. [CrossRef]

91. Yokoyama, T.; Toki, T.; Aoki, Y.; Kanezaki, R.; Park, M.J.; Kanno, Y.; Takahara, T.; Yamazaki, Y.; Ito, E.; Hayashi, Y.; et al. Identification of TRIB1 R107L gain-of-function mutation in human acute megakaryocytic leukemia. *Blood* **2012**, *119*, 2608–2611. [CrossRef]

92. Yamada, K.; Park, H.M.; Rigel, D.F.; DiPetrillo, K.; Whalen, E.J.; Anisowicz, A.; Beil, M.; Berstler, J.; Brocklehurst, C.E.; Burdick, D.A.; et al. Small-molecule WNK inhibition regulates cardiovascular and renal function. *Nat. Chem. Biol.* **2016**, *12*, 896–898. [CrossRef] [PubMed]

93. Burke, J.R.; Cheng, L.; Gillooly, K.M.; Strnad, J.; Zupa-Fernandez, A.; Catlett, I.M.; Zhang, Y.; Heimrich, E.M.; McIntyre, K.W.; Cunningham, M.D.; et al. Autoimmune pathways in mice and humans are blocked by pharmacological stabilization of the TYK2 pseudokinase domain. *Sci. Transl. Med.* **2019**, *11*, eaaw1736. [CrossRef] [PubMed]

94. Moslin, R.; Zhang, Y.; Wrobleski, S.T.; Lin, S.; Mertzman, M.; Spergel, S.; Tokarski, J.S.; Strnad, J.; Gillooly, K.; McIntyre, K.W.; et al. Identification of N-Methyl Nicotinamide and N-Methyl Pyridazine-3-Carboxamide Pseudokinase Domain Ligands as Highly Selective Allosteric Inhibitors of Tyrosine Kinase 2 (TYK2). *J. Med. Chem.* **2019**. [CrossRef] [PubMed]

95. Zhavoronkov, A.; Ivanenkov, Y.A.; Aliper, A.; Veselov, M.S.; Aladinskiy, V.A.; Aladinskaya, A.V.; Terentiev, V.A.; Polykovskiy, D.A.; Kuznetsov, M.D.; Asadulaev, A.; et al. Deep learning enables rapid identification of potent DDR1 kinase inhibitors. *Nat. Biotechnol.* **2019**, *37*, 1038–1040. [CrossRef] [PubMed]

Article

Bienzymatic Cascade for the Synthesis of an Optically Active O-benzoyl Cyanohydrin

Laura Leemans [1,2], Luuk van Langen [2], Frank Hollmann [3] and Anett Schallmey [1,*

1   Institute for Biochemistry, Biotechnology and Bioinformatics, Technische Universität Braunschweig, Spielmannstr. 7, 38106 Braunschweig, Germany; l.leemans-martin@tu-braunschweig.de
2   Viazym B.V., Molengraaffsingel 10, 2629 JD Delft, The Netherlands; vanlangen@viazym.nl
3   Department of Biotechnology, Delft University of Technology, Van der Maasweg 9, 2629 HZ Delft, The Netherlands; f.hollmann@tudelft.nl
*   Correspondence: a.schallmey@tu-braunschweig.de; Tel.: +49-531-391-55400

Received: 14 May 2019; Accepted: 10 June 2019; Published: 12 June 2019

Abstract: A concurrent bienzymatic cascade for the synthesis of optically pure (*S*)-4-methoxymandelonitrile benzoate ((*S*)-**3**) starting from 4-anisaldehyde (**1**) has been developed. The cascade involves an enantioselective *Manihot esculenta* hydroxynitrile lyase-catalyzed hydrocyanation of **1**, and the subsequent benzoylation of the resulting cyanohydrin (*S*)-**2** catalyzed by *Candida antarctica* lipase A in organic solvent. To accomplish this new direct synthesis of the protected enantiopure cyanohydrin, both enzymes were immobilized and each biocatalytic step was studied separately in search for a window of compatibility. In addition, potential cross-interactions between the two reactions were identified. Optimization of the cascade resulted in 81% conversion of the aldehyde to the corresponding benzoyl cyanohydrin with 98% enantiomeric excess.

Keywords: Enantioselectivity; enzyme cascade; hydroxynitrile lyase; lipase; hydrocyanation; transesterification

---

1. Introduction

In the past decades, the field of biocatalysis has grown exponentially, becoming an industrially attractive technology for the production of chiral pharmaceutical intermediates and precursors in the synthesis of fine chemicals [1,2]. Recently, considerable progress has been made in developing enzymatic cascades for the production of valuable chemical products [3–6]. These are synthetic processes that combine two or more enzymes in one pot and can be performed in a concurrent (simultaneous) or sequential mode [7]. The combination of individual reaction steps in a one-pot process has several advantages in terms of process efficiency and sustainability, since the solvent consumption and waste generation is generally decreased due to the lower number of work-up steps [5]. Furthermore, the coupling of reactions has the potential to drive equilibria towards the desired products, hence reducing the required excess of reagents [8]. There are, however, some challenges associated with the design of enzymatic cascades, mainly due to enzymes often having different optimum conditions. Furthermore, the presence of side products of certain reactions may lead to enzyme inhibition, which could be avoided by running the reactions in a sequential mode [9,10].

A current challenge in the field of enzymatic cascades is the coupling of a hydroxynitrile lyase (HNL)-catalyzed cyanohydrin synthesis with an acylation catalyzed by a lipase. The immediate acylation of the formed cyanohydrin could prevent the back reaction from taking place and yield a chemically more stable product. In addition, if a lipase with appropriate enantioselectivity is chosen, the enantiopurity of the final product can be enhanced.

However, the one-pot combination of the two enzymes is not trivial, for they have very different requirements. It is well known that water acts as a competing nucleophile in lipase-catalyzed transesterification reactions [11], whereas HNLs require relatively high water content when working in organic solvents [12–14]. Due to these different water content requirements, the cascade synthesis of acylated cyanohydrins catalyzed by an HNL and a lipase is a challenging task. Hanefeld et al. attempted to couple the hydrocyanation of benzaldehyde with the acetylation of the formed mandelonitrile, in order to shift the equilibrium of the first reaction. The reactions were catalyzed by *Hevea brasiliensis* hydroxynitrile lyase (*Hb*HNL) and *Candida antarctica* lipase B (CALB), respectively. However, they encountered problems due to hydrolysis of the acyl donor and subsequent deactivation of the *Hb*HNL [10]. As an alternative to the bienzymatic synthesis of acylated cyanohydrins, researchers have studied kinetic resolution of acylated cyanohydrins and dynamic kinetic resolution approaches combining an unselective hydrocyanation with an enantioselective transesterification [15–20]. Nevertheless, it has been shown that under certain conditions HNLs can work at low water content [21], opening the possibility for combining the two mentioned enantioselective enzymatic steps.

In the present study, we have selected the hydrocyanation of 4-anisaldehyde (**1**) combined with the benzoylation of the formed cyanohydrin (**2**) to yield 4-methoxymandelonitrile benzoate (**3**) as a model reaction (Scheme 1). The hydrocyanation of 4-anisaldehyde requires a particularly high excess of HCN to drive the equilibrium towards the cyanohydrin, compared to other benzaldehydes [22–24], which is attributed to the positive mesomeric effect of the *para*-methoxy substituent. The direct benzoylation of the cyanohydrin was envisioned as a possible solution to avoid isolating this instable product. The obtained cascade product **3** is interesting, as it can be transformed in a single hydrogenation step into (*S*)-tembamide [25], a naturally occurring *N*-(β-hydroxy) amide for which antiviral (HIV) and hypoglycemic activity have been reported [26,27].

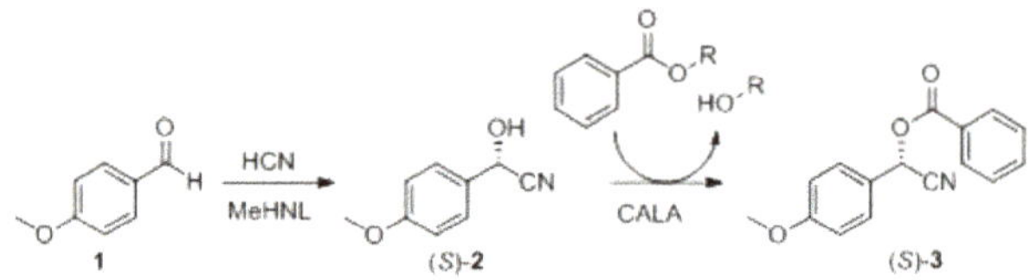

Scheme 1. Proposed cascade synthesis of (*S*)-4-methoxymandelonitrile benzoate.

Herein, *Manihot esculenta* hydroxynitrile lyase (*Me*HNL)—an *S*-selective enzyme [28] that can catalyze the hydrocyanation of 4-anisaldehyde—was studied. For the benzoylation of the resulting cyanohydrin, *Candida antarctica* lipase A (CALA) was chosen, which is known to accept bulky substrates and displayed preference for (*S*)-4-methoxymandelonitrile benzoate in a prior hydrolase screening (see supporting information Table S1 and Figure S1). Moreover, CALA is a robust industrial enzyme, and was shown to exhibit acyltransferase activity even in aqueous media [29], which we deemed an attractive feature in our envisioned cascade.

2. Results and Discussion

A crucial step in the development of the intended cascade was the identification of a suitable benzoyl donor that would facilitate the CALA-catalyzed transesterification. Several donors were screened in the CALA-catalyzed benzoylation of 4-methoxymandelonitrile (see Figure S2). Of the screened esters, phenyl benzoate afforded the highest reaction rate.

For a successful combination of *Me*HNL and CALA in the cascade synthesis of (*S*)-4-methoxymandelonitrile benzoate, the enzymatic hydrocyanation and transesterification reactions were subjected to a systematic study of the relevant reaction variables. The effects of temperature and water activity on both reactions were evaluated and a suitable carrier was selected for the immobilization of each enzyme. In a preliminary solvent screening, isopropyl ether was selected as reaction solvent due to its compatibility with both enzymes and the relatively high solubility of the substrates.

2.1. Influence of Reaction Temperature

In general, temperature affects enzymatic stability and has a great effect on the reaction rate of enzyme-catalyzed reactions and their enantioselectivity. The influence of temperature on the enantioselectivity of enzymatic reactions can be explained by a simple theoretical model [30]. Depending on whether the reaction takes place below or above the so-called racemic temperature, a decrease in temperature will increase or decrease the enantioselectivity, respectively.

According to literature, *MeHNL* shows optimal hydrocyanation activity at ~40 °C [31]. A reduction in reaction temperature below 40 °C would decrease enzymatic activity, but is expected to slightly increase its enantioselectivity, since *MeHNL* displays an elevated racemic temperature of 580 °C in the hydrocyanation of phenylpropionaldehyde in dibutylether [14]. Furthermore, decreasing the reaction temperature should reduce the rate of the nonenzymatic hydrocyanation of 4-anisaldehyde to a higher extent than the enzymatic reaction, which would also contribute to increase the enantiomeric excess (e.e.) of the cyanohydrin [32–34]. Finally, the reaction must be performed at a rather low temperature, since the boiling point of HCN is 25.6 °C [35]. Based on this information, hydrocyanation reactions were performed at 5, 10, and 20 °C.

Within the studied temperature range, the hydrocyanation rate increased moderately with the reaction temperature (see Figure 1). On the other hand, the enantioselectivity of the enzymatic reaction slightly increased at lower temperatures, as expected. These results suggest that the optimal reaction temperature, which would afford excellent enantioselectivity with a satisfactory reaction rate, would be 10 °C.

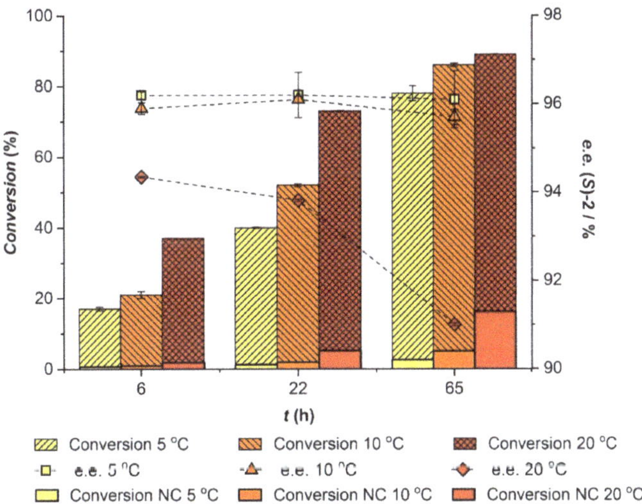

Figure 1. Effect of reaction temperature on the *MeHNL*-catalyzed hydrocyanation of 100 mM 4-anisaldehyde in a biphasic system using 6.5 equivalents of HCN. Negative control (NC) shows the rate of the unselective background reaction.

In the case of CALA, based on the data from kinetic resolution studies, the lipase was expected to show increased enantioselectivity with lower temperatures [36,37]. However, immobilized CALA has been reported to display a temperature optimum of 90 °C [38], thus the conversion will be limited by low reaction rates at low temperatures.

The effect of reaction temperature on the CALA-catalyzed benzoylation of racemic 4-methoxymandelonitrile was studied in isopropyl ether without added water in the range of 5 to 25 °C. The results showed that, when reducing the temperature, the enantioselectivity of CALA was slightly enhanced (E value of 7 at 5 and 10 °C and 6 at 25 °C), although this was at the expense of a pronounced reduction of reaction rate, as shown in Figure 2. Within the studied temperature range, the CALA-catalyzed benzoylation of 4-methoxymandelonitrile can be achieved in relatively high yields with optimal performance at 25 °C. Although CALA shows preference for conversion of the (*S*)-enantiomer, it is not selective enough to afford an enantiopure product. However, when preceded by an enantioselective hydrocyanation, the transesterification reaction can enhance the enantiopurity of the final product. Therefore, the observed minor effect of reaction temperature on the enantiomeric ratio can be neglected.

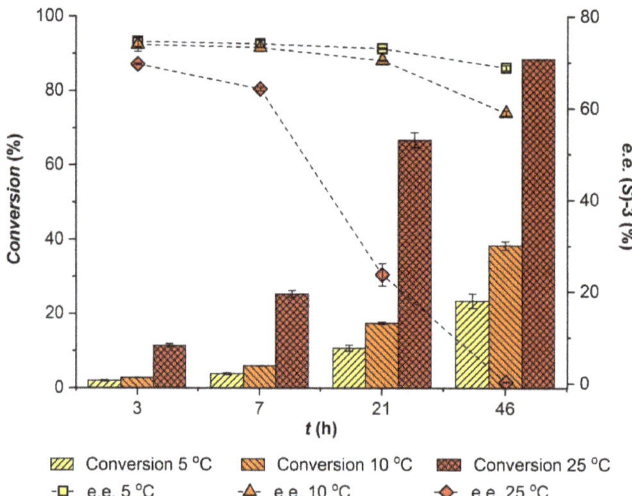

Figure 2. Effect of temperature on CALA-catalyzed benzoylation of 65 mM (±)-4-methoxymandelonitrile using 200 mM phenyl benzoate.

2.2. Selection of Enzyme Carriers

In our designed cascade, we envisioned a low-water organic solvent system for which the immobilization of both enzymes would be required. In organic solvents, free lipases generally express low catalytic activity and they tend to form aggregates which can lead to mass transfer limitations [39]. Adsorption is believed to enhance the catalytic activity of lipases due to hydrophobic interactions between the support and part of the enzyme yielding the active enzyme conformation by opening the lid [40]. Similarly, immobilization of hydroxynitrile lyases can improve their activity and stability towards organic solvents [41], and may lead to higher enantioselectivity [42]. In this study, CALA was adsorbed on Relizyme EXE309, a macroporous bead based on a methyl methacrylate/styrene copolymer that was recommended by the manufacturer for effective lipase adsorption. For *Me*HNL, adsorption on Celite R-633 was identified as the best immobilization method (see supporting information Table S2).

2.3. Influence of Water Activity

When working with enzymes in organic solvents, their activity is greatly affected by the water content in the reaction medium. This variable is best quantified in terms of thermodynamic water activity (a_w), which is equal in all phases in equilibrium [43]. Water is believed to act as a molecular lubricant, increasing the conformational flexibility of enzymes and, thus, their catalytic activity [44]. The water content also affects enzyme selectivity in a complex way, which involves interactions with the reaction medium and the substrates [14].

In the case of HNLs it has been shown that when working at low water activities, the enzymes are insufficiently hydrated, resulting in activity and selectivity loss [12–14]. Lipases, on the other hand, can generally withstand lower water content [15], and some studies have shown that a_w does not significantly influence the enantioselectivity of lipases [14,45–47]. Furthermore, water competes with the substrate alcohol (in this case, cyanohydrin) for the nucleophilic attack of the acyl-enzyme intermediate, leading to ester hydrolysis [48].

A prerequisite for the combination of both enzymes in a concurrent mode is to find a suitable a_w range that allows for sufficient *Me*HNL activity and high selectivity, while minimizing ester hydrolysis catalyzed by CALA. In order to evaluate the a_w effect on both enzymatic reactions, a salt hydrate system that could cover a wide range of a_w values was selected: $Na_2HPO_4/Na_2HPO_4·2H_2O/Na_2HPO_4·7H_2O/Na_2HPO_4·12H_2O$. These salt pairs afford water activity values of 0.15, 0.57 and 0.74, respectively, at 20 °C [43]. The hydrocyanation and benzoylation reactions were studied independently using immobilized *Me*HNL and CALA and the three salt pairs.

Before evaluating the influence of a_w on *Me*HNL, the effect of the phosphate salt hydrate pairs on the background hydrocyanation reaction was studied. The chemical hydrocyanation of 100 mM 4-anisaldehyde in isopropyl ether was performed using 6.5 equivalents of HCN and 0.5 mmol salt pair per mL of reaction. This revealed that the $Na_2HPO_4/Na_2HPO_4·2H_2O$ pair could convert 32% of the aldehyde in 24 h, whereas the other two salt pairs afforded a much lower chemical background reaction rate with only 4% conversion after 24 h. The negative control, where no salt was added, showed 1% conversion.

The effect of the three water activity values provided by the phosphate salts on the *Me*HNL-catalyzed hydrocyanation of 4-anisaldehyde was then evaluated (see Figure 3). Results at $a_w = 0.15$ are difficult to interpret due to the above-mentioned catalytic effect of the salt. It seems, however, that the enzyme is hardly active at this low water activity. Comparison of the results at the two highest water activity values, where the background reaction is similarly low, shows that the activity and selectivity of *Me*HNL increased with increasing a_w. Hence, to ensure a high e.e. of the resulting cyanohydrin, it is necessary to work at a high a_w for the combination of *Me*HNL and CALA.

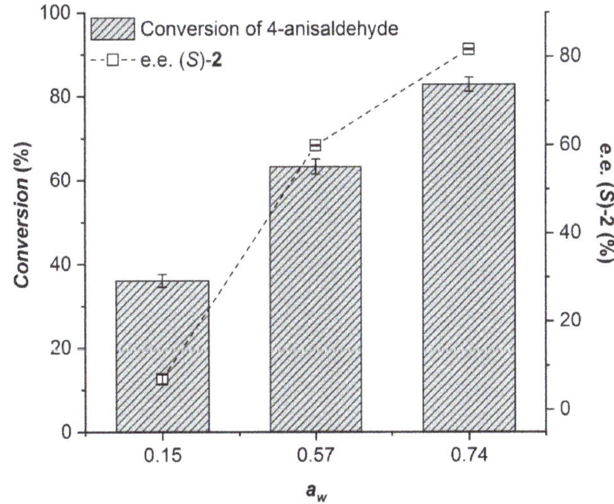

Figure 3. Effect of water activity on the *Me*HNL-catalyzed hydrocyanation of 90 mM 4-anisaldehyde using 6.5 equivalents of HCN. Conversion and enantiomeric excess (e.e.) values taken after 23 h of reaction.

In the case of CALA, at $a_w = 0.57$ and below, no significant difference in the enantioselectivity of the enzyme (E = 5) or in the selectivity for transesterification over hydrolysis was observed (see Figure 4). When working at $a_w = 0.74$, however, the hydrolysis rate increased significantly at the expense of the transesterification reaction. Although the enantioselectivity also increased under these conditions (E = 9), it is not a sufficient improvement to compensate for the increased hydrolysis rate. Considering the observed effects of water activity on *Me*HNL and CALA, we concluded that it should be possible to combine both enzymatic reactions at a_w = 0.57.

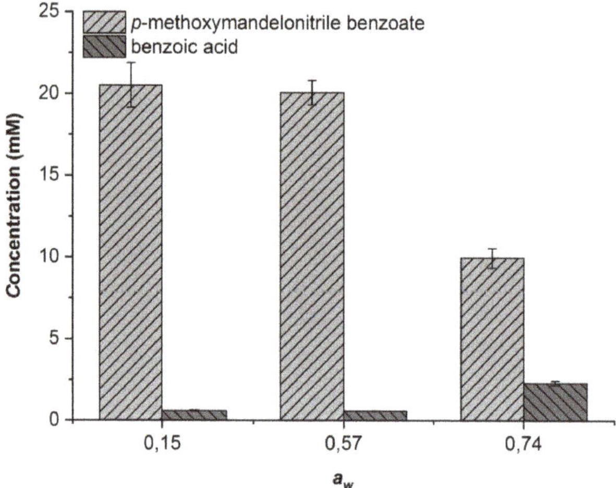

Figure 4. Effect of the different phosphate hydrate pairs on the CALA-catalyzed benzoylation of 66 mM (±)-4-methoxymandelonitrile using 200 mM vinyl benzoate after 22 h of reaction.

2.4. Cross Interactions

When performing reactions in one-pot, the reagents, products, or byproducts of one reaction may affect the other reactions. Thus, for an optimized cascade, the effect of hydrocyanation reagents on the CALA-catalyzed transesterification reaction as well as the influence of the benzoyl donor and the transesterification byproduct on the *Me*HNL-catalyzed hydrocyanation reaction was evaluated. Further, the influence of the side product benzoic acid on both transesterification and hydrocyanation reactions was studied.

The influence of 4-anisaldehyde and HCN on the reaction rate and selectivity of the CALA-catalyzed benzoylation of 4-methoxymandelonitrile was evaluated. As shown in Figure 5A, both 4-anisaldehyde and HCN exert a moderate negative effect on the benzoylation rate. Moreover, both compounds reduce the selectivity of CALA towards transesterification (see Figure 5B). Regarding the enantioselectivity of the benzoylation reaction, it was not significantly affected by the presence of either of the hydrocyanation substrates, with the E values varying between 5 and 6. Similarly, the effect of benzoic acid on CALA was studied in a benzoylation reaction of 70 mM (±)-4-methoxymandelonitirile containing one equivalent of benzoic acid. The reaction rate was 20% slower than that of the control.

The influence of phenyl benzoate and its leaving group, phenol, on the *Me*HNL-catalyzed hydrocyanation of 4-methoxymandelonitrile was studied. Addition of either three equivalents of phenyl benzoate or phenol to the hydrocyanation reaction afforded (*S*)-4-methoxymandelonitrile with the same yield and enantiopurity as the control reaction (data not shown). To evaluate the effect of the side product benzoic acid on the hydrocyanation reaction, an experiment was performed with addition of different amounts of the acid to the reaction mixture. As shown in Figure 6, benzoic acid exerts a strong negative effect on the hydrocyanation rate of *Me*HNL. Thus, in order to achieve a cascade synthesis of **3** with high yield and selectivity, the CALA-catalyzed hydrolysis of phenyl benzoate and **3** should be minimized as much as possible. This observation is not surprising, since benzoic acid has previously been reported as a competitive inhibitor for other hydroxynitrile lyases [49–51]. Based on these results, the observed negative effect of benzoic acid on CALA is minor when compared to its effect on *Me*HNL.

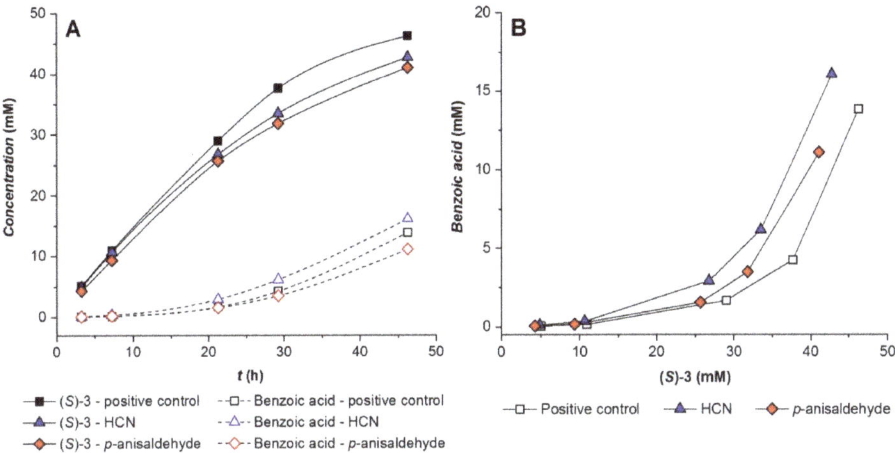

Figure 5. Effect of 60 mM 4-anisaldehyde and 209 mM HCN on the transesterification and hydrolysis rate of CALA (**A**) as well as its selectivity towards transesterification over hydrolysis (**B**) using 65 mM (±)-4-methoxymandelonitrile and 200 mM phenyl benzoate.

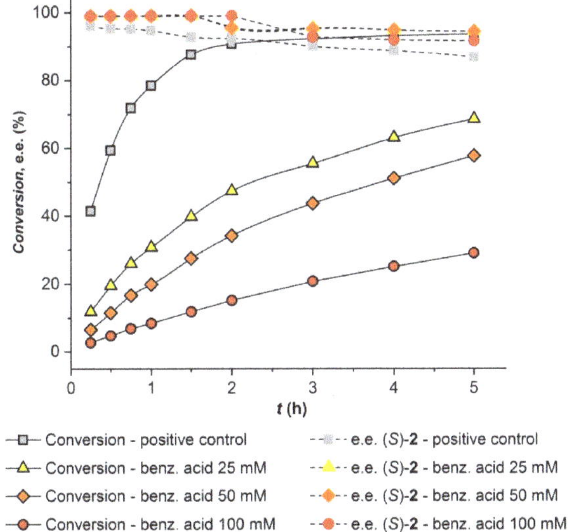

Figure 6. Effect of 25 mM, 50 mM, or 100 mM benzoic acid on *Me*HNL-catalyzed hydrocyanation of 100 mM 4-anisaldehyde using 6.5 equivalents of HCN. With exception of the positive control, the enantiomeric excess values overlap at low reaction times.

2.5. Cascade Synthesis of (S)-4-Methoxymandelonitrile Benzoate

For the first cascade synthesis test (see Table 1, entry 1), the used reaction conditions in terms of enzyme amount, temperature, water buffering salts and the time point of CALA addition were selected based on the results obtained before. The reaction temperature was set at 10 °C, $Na_2HPO_4 \cdot 2H_2O/Na_2HPO_4 \cdot 7H_2O$ was added to set the water activity to $a_w = 0.57$, and the lipase was added after 16 h of hydrocyanation reaction with the idea to minimize the exposure of *Me*HNL to benzoic acid. In this first reaction, a low reaction rate of the overall process was observed, resulting in only 10.5% conversion to the corresponding ester with a fair e.e. of 89% after 47 h reaction time.

The results, however, showed that both enzymes were simultaneously active after adding the lipase to the reaction mixture, since the 4-anisaldehyde was further converted and (*S*)-4-methoxymandelonitrile benzoate was also formed. The same process was repeated at 25 °C (Table 1, entry 2) resulting in a significantly improved reaction rate, while maintaining the enantiomeric excess of the ester. Based on these observations, all further reactions were performed at room temperature (20 °C).

For comparison, in experiment 3 (Table 1, entry 3) a concurrent approach, where both enzymes were added from the beginning, was tested. Due to the inhibitory effect of 4-anisaldehyde and HCN on the acyltransferase activity of CALA, a higher amount of lipase was added. This approach significantly improved the enantiopurity of the final ester (97% e.e.), while maintaining a good reaction rate. After 30.5 h of reaction, 10 mM cyanohydrin was still present in the reaction mixture. This suggested that the transesterification rate was not too high in comparison to the *Me*HNL-catalyzed hydrocyanation. Therefore, it was decided to maintain the CALA/*Me*HNL ratio for further reactions. Strikingly, the results could be further improved by omitting the water buffering salts from the system (Table 1, entry 4). Elimination of these salts provided overall higher yield and selectivity, which may be explained by the previously identified negative effect of low water activity on the enzymatic hydrocyanation reaction. Moreover, comparison of the benzoic acid concentration in experiments 3 and 4 indicated that in the former cascade (including the salts) the hydrolysis reaction was not suppressed to a higher extent than in the latter (without salts). Thus, the use of the salt hydrate pair did not further reduce the hydrolysis reaction in the studied system.

Salt-free experiments with varying initial amounts of benzoyl donor (Table 1, entries 4–6) showed a detrimental effect of higher donor concentrations on the hydrocyanation reaction, while the amount of hydrolysis side product, benzoic acid, increased substantially. In all cases, however, excellent enantiopurity of (*S*)-4-methoxymandelonitrile benzoate (99% e.e.) was maintained throughout the process. These results prompted us to test gradual addition of phenyl benzoate to the process, which was found to minimize the hydrolytic side reaction at the expense of some loss in reaction rate (see Table 1, entry 7), leading to 64% yield of ester **3** (98% e.e.) in 49 h, with 50% less benzoic acid produced compared to Table 1, entry 5. Finally, we hypothesized that a reduction of the lipase units could further decrease the amount of formed benzoic acid. Using 0.5 U/ml CALA afforded ester **3** (98% e.e.) with 81% yield in 122 h, with a total conversion of 4-anisaldehyde of 95% (see Table 1, entry 8 and Figure S3). The reaction time required for high conversion values could be reduced by approximately one day when the reaction temperature was increased from 20 to 25 °C (see Table 1, entry 10).

Furthermore, we also tested the possibility of decreasing the amount of HCN used in the cascade. However, when performing the cascade using 5 instead of 6.5 equivalents of HCN (using the same conditions detailed in Table 1, entry 7), after 122 h only 60 mM (*S*)-**3** was produced with a poor enantiomeric excess of 87%. Similarly, a decrease in yield and enantiomeric excess of (*S*)-**2** was observed in the immobilized *Me*HNL-catalyzed hydrocyanation of 4-anisaldehyde when decreasing the HCN concentration (see Figure S4).

Overall, our results demonstrate that the concurrent combination of a *Me*HNL-catalyzed hydrocyanation with a CALA-catalyzed transesterification is possible after careful investigation of the individual reactions in search for compatible conditions. However, further optimization is required to improve the cascade, since 5% residual 4-anisaldehyde could not be converted by *Me*HNL (see Table 1, entry 8 and Figure S4A). Presumably, this is due to an observed increase in benzoic acid concentration as the cascade proceeds. Moreover, roughly 11 mM cyanohydrin **2** remained in the reaction media, which could not be further converted to **3** by CALA. This indicates a rather low affinity of the lipase for 4-methoxymandelonitrile, which, at low concentrations, cannot compete with the water present in the reaction medium for the acyl-enzyme complex. Hence, hydrolysis is the main reaction taking place in the end, further increasing the amount of benzoic acid.

The use of an engineered CALA with enhanced selectivity towards transesterification reactions versus hydrolysis could significantly improve the performance of our designed cascade [29].

Table 1. Cascade synthesis of (S)-4-methoxymandelonitrile benzoate catalyzed by immobilized *Me*HNL and CALA starting from 100 mM 4-anisaldehyde and 650 mM HCN using 200–400 mM phenyl benzoate. Reactions were performed in duplicate and relative standard deviations were generally below 2%.

Exp.	T (°C)	Equiv. Phenyl Benzoate	*Me*HNL (U/mL) [1]	CALA (U/mL) [2]	Time (h)	[1] (mM)	[2] (mM)	e.e. (S)-2 (%)	[3] (mM)	e.e. (S)-3 (%)	[benzoic acid] (mM)
1 [3]	10	3	0.21	0.50 [4]	16	90	10	58	-	-	-
					47	82	7	12	11	89	ND
2 [3]	25	3	0.21	0.50 [5]	15	75	25	63	-	-	-
					24	59	11	15	30	90	ND
3 [3]	20	2	0.26	2.20	31	47	10	70	43	97	64
4	20	2	0.26	2.20	31	23	21	92	56	99	57
					73	15	23	86	62	98	111
5	20	3	0.26	2.20	31	33	8	76	59	99	109
6	20	4	0.26	2.20	31	45	4	68	51	99	140
7	20	3 [6]	0.26	2.20	49	18	19	74	64	98	56
					122	15	13	79	72	93	210
8	20	3 [6]	0.26	0.50	122	5	14	60	81	98	50
					164	4	11	61	85	96	94
9	20	0	0.26	0	24	15	85	91	-	-	-
10	25	3 [3]	0.26	0.50	94	7	13	46	80	97	39

[1] Activity of immobilized *Me*HNL: 0.03 U/mg [2] Activity of immobilized CALA: 0.11 U/mg. [3] $Na_2HPO_4 \cdot 2H_2O$/$Na_2HPO_4 \cdot 7H_2O$ was added (0.34 mmol per mL of reaction). [4] CALA was added after 16 h. [5] CALA was added after 15 h. [6] Gradual addition: Each 1 equivalent of phenyl benzoate was added at the beginning, after 24 h, and after 48 h.

3. Materials and Methods

3.1. Enzymes

Hydroxynitrile lyase from *Manihot esculenta* was heterologously expressed in *E. coli* K12 Top 10F′ harboring the pSE420-*Me*HNL plasmid, which was kindly provided by Dr. Kerstin Steiner from the Austrian Centre of Industrial Biotechnology (ACIB). 2 × YT medium (100 mL, containing 100 µg/mL ampicillin) was inoculated from a glycerol stock (10 µL) and the culture was incubated overnight at 37 °C, 200 rpm. A 5 mL aliquot of this preculture (OD_{600} = 4.4) was used to inoculate 2 × YT medium (500 mL, containing 100 µg/mL ampicillin), and the new culture was grown at 37 °C, 200 rpm. At an optical density at 600 nm (OD_{600}) of 0.8, IPTG was added to a final concentration of 0.1 mM and the culture was further incubated at 18 °C, 160 rpm for 48 h. Cells were harvested by centrifugation at 4000 g, 4 °C for 10 min, washed with 25 mM sodium acetate buffer, pH 5.8, and centrifuged again (4000 g, 4 °C for 30 min). Cell pellets were resuspended in 25 mM sodium acetate, pH 5.8 (4 mL per g of wet cells), and 10 mL aliquots were sonicated at 4 °C using a Fisher Scientific Model 120 Sonic Dismembrator at 60% amplitude, 2 s pulse, 5 s pause for 7 min. The resulting crude extract was centrifuged to remove cell debris (18,500 g, 4 °C for 45 min) and the obtained cell free extract was stored at −20 °C. Concentration of cell free extract was performed in an Amicon Ultra Filter Device with 10 kDa cut-off at 4 °C.

Candida antarctica lipase A (Novozym®CALA L, LDN 00025) was purchased from Novozymes.

3.1.1. Enzyme Activity Measurements

The enzymatic activity of free *Me*HNL was measured following the cleavage of racemic mandelonitrile into benzaldehyde and HCN, according to a literature procedure [52]. Samples were prepared by diluting the *Me*HNL cell free extract in 5 mM phosphate buffer, pH 6.5. The enzymatic activity of immobilized *Me*HNL was determined following the hydrocyanation of 4-anisaldehyde (100 mM) using 6.5 equivalents of HCN in isopropyl ether at 20 °C, 800 rpm. A negative control was performed using the enzyme support instead. Samples were analyzed by chiral HPLC.

The enzymatic activity of CALA was measured according to a modified literature procedure following the hydrolysis of tributyrin in phosphate buffer at pH 7 [53]. The resulting butyric acid was titrated with 0.1 M sodium hydroxide, and the consumption of the latter was recorded as a function of time. The enzymatic activity of immobilized CALA was determined following the hydrolysis of 4-nitrophenyl butyrate (1.67 mM) in acetonitrile/50 mM phosphate pH 7.4 (1/5). A calibration curve of the product 4-nitrophenol was used to calculate the reaction rate based on the absorbance of the reaction mixture at 410 nm.

3.1.2. Enzyme Immobilization

General procedure for *Me*HNL: In a 5 mL glass vial containing a magnetic stirring bar and the carrier (100 mg), *Me*HNL-containing cell-free extract (40 µL) was added dropwise while stirring. The mixture was stirred for 30 min and was then dried at room temperature under high vacuum for 48 h.

Immobilization of CALA: CALA (30 g), Relizyme EXE309 (30 g), and demineralized water (30 mL) were mixed in a 100 mL bottle. The pH was adjusted to 6 with a solution of 2 M sodium hydroxide and the mixture was incubated at room temperature in an orbital shaker overnight. After filtering the solution, the activity of the supernatant was measured and compared to the initial activity of the enzyme solution before immobilization, giving an immobilization yield of 84%. The immobilized enzyme was washed several times with demineralized water and dried at room temperature under high vacuum for 48 h.

3.1.3. Evaluation of the *Me*HNL Immobilisates

To 1.5 mL glass vials containing different *Me*HNL immobilisates (15 mg) and $Na_2HPO_4 \cdot 2H_2O$ (30 mg, 0.17 mmol), each a solution of 100 mM 4-anisaldehyde and 650 mM HCN in isopropyl ether (500 µL) was added. Duplicate reactions per immobilisate were incubated at 20 °C in an orbital shaker at 800 rpm. Samples (20 µL) were analyzed by HPLC.

3.2. Chemicals

Unless otherwise indicated, reagents and organic solvents were purchased from Fisher Scientific, Sigma-Aldrich, TCI Chemicals and Acros Organics, and were of the highest available purity.

Celite R633 was a gift from Imerys S.A., Relizyme EXE309 was kindly provided by Resindion S.r.l., and SYLOID silicas were a gift from Grace Davison, Inc. (±)-4-Methoxymandelonitrile was synthetized via chemical hydrocyanation of 4-anisaldehyde following a literature procedure [22].

For the preparation of hydrogen cyanide solution, potassium cyanide (26 g, 0.4 mol) was dissolved in a stirred mixture of distilled water (200 mL) and diisopropyl ether (^{i}Pr$_2$O) (100 mL). The solution was cooled in an ice-water bath and 1.33 M citric acid (100 mL, 0.13 mol, 1 equivalent) was slowly added. The aqueous layer was extracted twice with 50 mL of ^{i}Pr$_2$O. The combined organic phases were stored in a dark bottle containing 1 M citrate buffer pH 5.5 (10 mL). The HCN concentration was determined by measuring the absorption of the $[Ni(CN)_4]^{2-}$ complex ion at 267 nm. HCN solution (20 µL) was added to 3.5 mM $NiSO_4 \cdot 6H_2O$ in 1 M NH_3 (4 mL) [54]. After vigorous mixing, the solution was further diluted 50 times with 1 M NH_3 in a cuvette. The absorption was measured and the resulting HCN concentration was calculated based on a calibration curve.

Synthesis of (±)-4-Methoxymandelonitrile (**2**)

In a 100 mL round-bottom flask, NaCN (4.4 g, 90 mmol) was dissolved in water (30 mL) and NaHSO$_3$ (9.4 g, 90 mmol) was slowly added. 4-Anisaldehyde (2.4 g, 18 mmol) was dissolved in ethyl acetate (20 mL) and added to the reaction mixture. After 1 h of vigorous stirring, the flask was introduced in an ice-water bath and stirred for another 6 h. The reaction was followed by HPLC until it reached plateau after 90% conversion. Distilled water (30 mL) was added to dissolve the salts and the aqueous phase was extracted ethyl acetate (2 × 10 mL). The combined organic phase was washed with brine, dried over anhydrous MgSO$_4$, and concentrated on a rotary evaporator using a 10 °C bath, whereupon crystallization occurred. The crystals were washed with cold heptane/ethyl acetate (2:1) and stored at 5 °C. HPLC analysis showed that the (±)-4-methoxymandelonitrile crystals contained 3% 4-anisaldehyde impurity.

3.3. Bioconversions

All bioconversions were performed in 1.5 mL glass vials in an orbital shaker at 800 rpm. Unless otherwise stated, the reaction temperature was set at 20 °C.

3.3.1. Benzoyl Donor Screening

To a vial containing immobilized CALA (4.2 mg, 0.46 U), racemic 4-methoxymandelonitrile (200 mM) in isopropyl ether (250 µL) and benzoyl donor (300 mM) in isopropyl ether (500 µL) was added. Reactions were performed in duplicate and were incubated at 25 °C. Samples (20 µL) were taken after 4, 20, and 27.5 h and analyzed by HPLC. The enantioselectivity was calculated according to the method of Chen et al. for conversion values between 5 and 20% [55].

3.3.2. Temperature Effect on *Me*HNL

4-Anisaldehyde (530 mM) in isopropyl ether (188 µL), HCN (800 mM) in isopropyl ether (812 µL), citrate buffer pH 5 (500 mM, 125 µL), and *Me*HNL-containing cell-free extract (33.2 µL, 5 U) were incubated at 5, 10, and 20 °C. Reactions were performed in duplicate and a negative control reaction was performed for each temperature value, where no enzyme was added. Samples (20 µL) were taken from the upper organic phase after 6, 22, and 75 h and analyzed by HPLC.

3.3.3. Temperature Effect on CALA

To a vial containing CALA (2 mg, 0.22 U), (±)-4-methoxymandelonitrile (195 mM) in isopropyl ether (125 µL) and phenyl benzoate (300 mM) in isopropyl ether (250 µL) was added. The mixture was incubated at 5, 10, and 25 °C. Reactions were performed in duplicate. Samples (20 µL) were taken after 3, 7, 21, and 46 h and analyzed by HPLC.

3.3.4. Effect of Na_2HPO_4 Salt Hydrates on Chemical Hydrocyanation

To a solution of 4-anisaldehyde (100 mM) and HCN (650 mM) in isopropyl ether (1 mL), anhydrous Na_2HPO_4 (70 mg, 0.5 mmol), $Na_2HPO_4 \cdot 2H_2O$ (85 mg, 0.5 mmol), or $Na_2HPO_4 \cdot 7H_2O$ (134 mg, 0.5 mmol) was added. Reactions were performed in duplicate and a control reaction without salt was run in parallel. The amount of salt added is based on the maximum solubility of water in isopropyl ether (0.55 wt%) [56] and the water capacity of the $Na_2HPO_4/Na_2HPO_4 \cdot 2H_2O$ salt pair (11.2 mmol/g) [57]. Reactions were stopped after 24 h and analyzed by HPLC.

3.3.5. Effect of a_w on *Me*HNL

To a vial containing *Me*HNL immobilized on Celite R633 (12 mg, 0.36 U), a solution of 4-anisaldehyde (90 mM) and HCN (590 mM) in isopropyl ether (550 µL) was added, together with either anhydrous Na_2HPO_4 (35 mg, 0.25 mmol), $Na_2HPO_4 \cdot 2H_2O$ (42 mg, 0.25 mmol) or $Na_2HPO_4 \cdot 7H_2O$ (67 mg, 0.25 mmol). Reactions were performed in duplicate. Samples (20 µL) were taken after 5 and 24 h and analyzed by HPLC.

3.3.6. Effect of a_w on CALA

To a vial containing CALA immobilized on Relizyme EXE309 (4 mg, 0.44 U), a solution of (±)-4-methoxymandelonitrile (100 mM) and vinyl benzoate (300 mM) in isopropyl ether (500 µL) was added, together with either anhydrous Na_2HPO_4 (35 mg, 0.25 mmol), $Na_2HPO_4 \cdot 2H_2O$ (42 mg, 0.25 mmol), or $Na_2HPO_4 \cdot 7H_2O$ (67 mg, 0.25 mmol). Samples (20 µL) were taken after 6, 22, and 30 h and analyzed by HPLC.

3.3.7. Effect of Phenol and Phenyl Benzoate on *Me*HNL

To a solution of 4-anisaldehyde (100 mM) and HCN (600 mM) in isopropyl ether (670 µL), *Me*HNL on Celite R633 (8 mg, 0.24 U) and either phenol (18.8 mg, 200 µmol) or phenyl benzoate (39.6 mg, 200 µmol) were added. A control was performed without addition of phenol or phenyl benzoate. Samples (20 µL) were taken after 17 and 22 h and analyzed by HPLC.

3.3.8. Effect of 4-Anisaldehyde and HCN on CALA

To a solution of (±)-4-methoxymandelonitrile (65 mM) and phenyl benzoate (200 mM) in isopropyl ether (375 µL), CALA on Relizyme EXE309 (2 mg, 0.22 U) and either 4-anisaldehyde (200 mM) in isopropyl ether (160 µL) or HCN (700 mM) in isopropyl ether (160 µL) were added. Reactions were performed in duplicate and a control was performed adding isopropyl ether (160 µL) instead of the hydrocyanation reagents. The reactions were incubated at 25 °C. Samples (20 µL) were taken after 3, 7, 21, 29, and 46 h and analyzed by HPLC.

3.3.9. Effect of Benzoic Acid on *Me*HNL

To a vial containing *Me*HNL on Celite R633 (15 mg, 0.45 U) and benzoic acid (1.5 mg, 12 µmol; 3 mg, 25 µmol; or 6 mg, 50 µmol), a solution of 4-anisaldehyde (100 mM), HCN (650 mM), and mesitylene

(100 mM) (internal standard) in isopropyl ether (500 µL) was added. A control was performed without addition of benzoic acid. Samples (20 µL) were analyzed by HPLC.

3.3.10. Effect of Benzoic Acid on CALA

To a solution of (±)-4-methoxymandelonitrile (100 mM), phenyl benzoate (300 mM) and mesitylene (100 mM) (internal standard) in isopropyl ether (350 µL), CALA on Relizyme EXE309 (2 mg, 0.22 U), and $Na_2HPO_4 \cdot 2H_2O$ (30 mg, 0.17 mmol), benzoic acid (70 mM) in isopropyl ether (150 µL, 1 equivalent) were added. A control was performed adding isopropyl ether (150 µL) instead of benzoic acid solution. Samples (20 µL) were analyzed by HPLC.

3.3.11. General Procedure for the Cascade Synthesis of (*S*)-4-Methoxymandelonitrile Benzoate

To a vial containing immobilized *Me*HNL and CALA, $Na_2HPO_4 \cdot 2H_2O$ (30 mg, for experiments 1-3, Table 1) was added. A solution of 4-anisaldehyde (360 mM) and mesitylene (360 mM) (internal standard) in isopropyl ether (140 µL) and HCN (900 mM) in isopropyl ether (360 µL) was subsequently added. To this mixture, phenyl benzoate (2-4 equivalents) was added (in case of experiments 7 and 8 from Table 1, 1 equivalent was added each time at t = 0, after 24 h and after 46 h) and the reactions were incubated at 10, 20, or 25 °C. Negative control reactions without enzyme addition were performed in parallel. Reactions were performed in duplicate and 20 µL samples were analyzed by HPLC.

3.4. HPLC Analysis

The quantification of 4-anisaldehyde, (*S*)- and (*R*)-4-methoxymandelonitrile and (*S*)- and (*R*)-4-methoxymandelonitrile benzoate was achieved using chiral normal phase HPLC. Samples from reaction mixtures were diluted 50× with n-heptane/isopropanol 8:2 v/v and dried over anhydrous magnesium sulfate. Analysis was performed using a Hitachi Elite LaChrom HPLC system, consisting of a VWR Hitachi L-2130 pump, L-2200 auto-sampler, L-2350 column oven, and L-2400 UV detector using a D_2 lamp coupled to a Hitachi organizer module. Separation was achieved on a Chiralpak AD-H column (250 mm × 4.6 mm × 5 µm; Daicel, Japan), using n-heptane/isopropanol (82/18) at a flow rate of 0.65 mL/min. The column oven temperature was set at 35 °C and a detection wavelength of 225 nm was chosen. Under these conditions, the compounds were separated with the following retention times; mesitylene (internal standard) 5.0 min, benzoic acid 6.7 min, phenol 7.0 min, phenyl benzoate 7.3 min, 4-anisaldehyde 8.3 min, (*S*)-4-methoxymandelonitrile 9.6 min, (*R*)-4-methoxymandelonitrile 10.5 min, (*S*)-4-methoxymandelonitrile benzoate 13.1 min, and (*R*)-4-methoxymandelonitrile benzoate 13.9 min.

Reverse-phase HPLC was used for the quantification of benzoic acid. Samples from reaction mixtures were diluted 50 × with acetonitrile/water 1:1 v/v. Analysis was performed using a Shimadzu Nexera XR HPLC system equipped with a DGU-20A5R degassing unit, coupled to two LC-20AD solvent delivery units, a SIL-20AC Autosampler, a CTO-20A column oven, an SPD-20A UV/VIS detector using a D_2 lamp, and a CBM-20A system controller. Separation was achieved on a Nucleoshell RP 18 column (150 mm × 3 mm × 2.7 µm; Macherey-Nagel, Germany) at 30 °C, 0.7 mL/min flow rate using MilliQ®water containing 0.1% trifluoroacetic acid (solvent A) and acetonitrile (solvent B) as mobile phase and UV detection at 225 nm. The following eluent program was used; 10% solvent B for 0.5 min, followed by a linear gradient to 45% solvent B in 5.5 min, 45% solvent B for 10 min, another linear gradient to 100% solvent B in 5 min, 100% solvent B for 3 min, and finally a linear gradient to 10% solvent B in 3 min followed by 10% solvent B for 3 min. Under these conditions, the compounds were separated with the following retention times: phenol 4.3 min, benzoic acid 4.9 min, (±)-4-methoxymandelonitrile 5.2 min, 4-anisaldehyde 5.9 min, phenyl benzoate 14.8 min, (±)-4-methoxymandelonitrile benzoate 16.1 min, and mesitylene (internal standard) 19.3 min.

Calibration curves were prepared using commercial 4-anisaldehyde (>99.0%), (*R*)-4-methoxymandelonitrile (98.0%) and benzoic acid (>99.5%) as well as chemically synthesized (±)-4-methoxymandelonitrile benzoate.

4. Conclusions

In the present work, the successful concurrent cascade synthesis of (*S*)-4-methoxymandelonitrile benzoate starting from 4-anisaldehyde and catalyzed by *Me*HNL and CALA was accomplished. Systematically studying each enzymatic reaction first was key for their effective combination. When developing the one-pot cascade, however, further fine-tuning of reaction conditions was required to achieve optimal performance. This way, 95% conversion of **1** with 81% yield of the final ester (*S*)-**3** and excellent enantiopurity of 98% e.e. was achieved. Compared to the enzymatic hydrocyanation performed under the same conditions, which yielded **2** with 85% conversion and 91% e.e. (see Table 1, entry 9, and Figure S4B), this result proves that the CALA-catalyzed benzoylation of (*S*)-**2** coupled to the *Me*HNL-catalyzed hydrocyanation of **1** can shift the equilibrium of the latter by removal of the unstable intermediate. Nevertheless, there is still room for further improvement, since the low affinity of CALA for cyanohydrin **2** has been identified as the main bottleneck that prevents the cascade from reaching full conversion.

In principle, our approach reported herein can be transferred to the synthesis of other acylated cyanohydrins. In any case, a key requirement for success will be the application of a lipase with high selectivity for transesterification over hydrolysis in microaqueous media.

Supplementary Materials: The following are available online at http://www.mdpi.com/2073-4344/9/6/522/s1, Figure S1: Conversion and enantiomeric excess (e.e.) of the remaining 4-methoxymandelonitrile benzoate after 18 h of hydrolysis reaction, Figure S2: Conversion and enantiomeric excess (e.e.) values obtained for the CALA-catalyzed benzoylation of 67 mM (±)-4-methoxymandelonitrile using 200 mM vinyl benzoate (VB), acetoxime benzoate (AB) or phenyl benzoate (PB), Figure S3: Comparison of the cascade synthesis of (S)- 4-methoxymandelonitrile benzoate catalyzed by *Me*HNL and CALA starting from 100 mM 1 (A, Table S2, entry 8) and the hydrocyanation of 100 mM **1** catalyzed by *Me*HNL under the same conditions (B, Table S2, entry 9), Figure S4: Hydrocyanation of 100 mM 4-anisaldehyde catalyzed by immobilized *Me*HNL using 3-6.5 equivalents of HCN at 20 °C. Table S1: Assigned number and specifications of the lipases screened in the hydrolysis of (±)-4-methoxymandelonitrile benzoate, Table S2: Results of hydrocyanation of 100 mM 4-anisaldehyde with 6.5 equivalents of HCN catalyzed by immobilized *Me*HNL in isopropyl ether with 0.34 mmol $Na_2HPO_4 \cdot 2H_2O$/$Na_2HPO_4 \cdot 7H_2O$ per mL (water activity of 0.57). Detailed information and additional data on the screening of hydrolases for the selective hydrolysis of (±)-4-methoxymandelonitrile benzoate, the screening of benzoyl donors, the immobilization of *Me*HNL, the effect of HCN concentration on *Me*HNL-catalyzed hydrocyanation, the optimized cascade, and the chemical synthesis of (±)-4-methoxymandelonitrile benzoate.

Author Contributions: Conceptualization, L.v.L. and F.H.; methodology, A.S.; validation, L.L., A.S., and L.v.L.; formal analysis, L.L.; investigation, L.L.; resources, A.S., L.v.L. and F.H.; data curation, A.S. and L.v.L.; writing—original draft preparation, L.L. and L.v.L.; writing—review and editing, A.S. and F.H.; visualization, L.L.; supervision, A.S., L.v.L., and F.H..; project administration, A.S. and L.v.L.; funding acquisition, L.v.L., F.H., and A.S.

Funding: This project was funded by the European Union's Horizon 2020 MSCA ITN-EID program under grant agreement No 634200 (BIOCASCADES). This communication reflects only the beneficiary´s view and the European Commission is not responsible for any use that may be made of the information it contains.

Acknowledgments: We thank Antje Spieß (Institute of Biochemical Engineering, Technische Universität Braunschweig, Germany) for fruitful discussion and for providing an HPLC system for chiral analysis. L. Leemans also thanks Yvonne Goecke (Institute of Biochemical Engineering, Technische Universität Braunschweig, Germany) for technical support with the HPLC equipment. Further, Kerstin Steiner (ACIB, Graz, Austria) is acknowledged for the generous gift of *E. coli* K12 Top 10F' clone harboring the pSE420-*Me*HNL plasmid. Imerys S.A. and Resindion S.r.l. are acknowledged for providing Celite R633 and supplying Relizyme EXE309, respectively.

References

1. Choi, J.M.; Han, S.S.; Kim, H.S. Industrial Applications of Enzyme Biocatalysis: Current Status and Future Aspects. *Biotechnol. Adv.* **2015**, *33*, 1443–1454. [CrossRef] [PubMed]
2. Gröger, H.; Asano, Y. Introduction-Principles and Historical Landmarks of Enzyme Catalysis in Organic Synthesis. In *Enzyme Catalysis in Organic Synthesis*; Drauz, K., Gröger, H., May, O., Eds.; Wiley-VCH Verlag GmbH & Co. KGaA: Weinheim, Germany, 2012; pp. 1–42.

3. Sperl, J.M.; Sieber, V. Multienzyme Cascade Reactions—Status and Recent Advances. *ACS Catal.* **2018**, *8*, 2385–2396. [CrossRef]

4. France, S.P.; Hepworth, L.J.; Turner, N.J.; Flitsch, S.L. Constructing Biocatalytic Cascades: In Vitro and in Vivo Approaches to de Novo Multi-Enzyme Pathways. *ACS Catal.* **2017**, *7*, 710–724. [CrossRef]

5. Gröger, H.; Hummel, W. Combining the "two Worlds" of Chemocatalysis and Biocatalysis towards Multi-Step One-Pot Processes in Aqueous Media. *Curr. Opin. Chem. Biol.* **2014**, *19*, 171–179. [CrossRef] [PubMed]

6. Gröger, H.; Hummel, W. Chemoenzymatic Multistep One-Pot Processes. *Cascade Biocatal.* **2014**, 427–456. [CrossRef]

7. Schrittwieser, J.H.; Velikogne, S.; Hall, M.; Kroutil, W. Artificial Biocatalytic Linear Cascades for Preparation of Organic Molecules. *Chem. Rev.* **2018**, *118*, 270–348. [CrossRef] [PubMed]

8. Sheldon, R.A.; Arends, I.W.C.E.; Hanefeld, U. Process Integration and Cascade Catalysis. In *Green Chemistry and Catalysis*; Wiley-VCH Verlag GmbH & Co. KGaA: Weinheim, Germany, 2007; pp. 389–408.

9. Ricca, E.; Brucher, B.; Schrittwieser, J.H. Multi-Enzymatic Cascade Reactions: Overview and Perspectives. *Adv. Synth. Catal.* **2011**, *353*, 2239–2262. [CrossRef]

10. Hanefeld, U.; Straathof, A.J.J.; Heijnen, J.J. Enzymatic Formation and Esterification of (*S*)-Mandelonitrile. *J. Mol. Catal. B Enzym.* **2001**, *11*, 213–218. [CrossRef]

11. Valivety, R.H.; Halling, P.J.; Macrae, A.R. Water as a Competitive Inhibitor of Lipase-Catalysed Esterification in Organic Media. *Biotechnol. Lett.* **1993**, *15*, 1133–1138. [CrossRef]

12. Paravidino, M.; Sorgedrager, M.; Orru, R.A.A.; Hanefeld, U. Activity and Enantioselectivity of the Hydroxynitrile Lyase *Me*HNL in Dry Organic Solvents. *Chem. Eur. J.* **2010**, *16*, 7596–7604. [CrossRef]

13. Costes, D.; Wehtje, E.; Adlercreutz, P. Hydroxynitrile Lyase-Catalyzed Synthesis of Cyanohydrins in Organic Solvents Parameters Influencing Activity and Enantiospecificity. *Enzyme Microb. Technol.* **1999**, *25*, 384–391. [CrossRef]

14. Persson, M.; Costes, D.; Wehtje, E.; Adlercreutz, P. Effects of Solvent, Water Activity and Temperature on Lipase and Hydroxynitrile Lyase Enantioselectivity. *Enzyme Microb. Technol.* **2002**, *30*, 916–923. [CrossRef]

15. Li, Y.-X.; Straathof, A.J.J.; Hanefeld, U. Enantioselective Formation of Mandelonitrile Acetate—investigation of a Dynamic Kinetic Resolution. *Tetrahedron Asymmetry* **2002**, *13*, 739–743. [CrossRef]

16. Veum, L.; Kuster, M.; Telalovic, S.; Hanefeld, U.; Maschmeyer, T. Enantioselective Synthesis of Protected Cyanohydrins. *Eur. J. Org. Chem.* **2002**, *2002*, 1516–1522. [CrossRef]

17. Veum, L.; Hanefeld, U. Enantioselective Formation of Mandelonitrile Acetate: Investigation of a Dynamic Kinetic Resolution II. *Tetrahedron Asymmetry* **2004**, *15*, 3707–3709. [CrossRef]

18. Veum, L.; Hanefeld, U. Enantioselective Synthesis of Aliphatic Cyanohydrin Acetates. *Synlett* **2005**, *15*, 2382–2384. [CrossRef]

19. Hietanen, A.; Ekholm, F.S.; Leino, R.; Kanerva, L.T. Applying Biocatalysis to the Synthesis of Diastereomerically Enriched Cyanohydrin Mannosides. *Eur. J. Org. Chem.* **2010**, *2010*, 6974–6980. [CrossRef]

20. Hietanen, A.; Kanerva, L.T. One-Pot Oxidation-Hydrocyanation Sequence Coupled to Lipase-Catalyzed Diastereoresolution in the Chemoenzymatic Synthesis of Sugar Cyanohydrin Esters. *Eur. J. Org. Chem.* **2012**, *2012*, 2729–2737. [CrossRef]

21. Okrob, D.; Paravidino, M.; Orru, R.V.A.; Wiechert, W.; Hanefeld, U.; Pohl, M. Hydroxynitrile Lyase from *Arabidopsis thaliana*: Identification of Reaction Parameters for Enantiopure Cyanohydrin Synthesis by Pure and Immobilized Catalyst. *Adv. Synth. Catal.* **2011**, *353*, 2399–2408. [CrossRef]

22. Malona, J.A.; Cariou, K.; Spencer, W.T.; Frontier, A.J. Total Synthesis of (±)-Rocaglamide via Oxidation-Initiated Nazarov Cyclization. *J. Org. Chem.* **2012**, *77*, 1891–1908. [CrossRef]

23. Yildirim, D.; Tükel, S.S.; Alagöz, D. Crosslinked Enzyme Aggregates of Hydroxynitrile Lyase Partially Purified from *Prunus dulcis* Seeds and Its Application for the Synthesis of Enantiopure Cyanohydrins. *Biotechnol. Prog.* **2014**, *30*, 818–827. [CrossRef] [PubMed]

24. Van Langen, L.M.; Van Rantwijk, F.; Sheldon, R.A. Enzymatic Hydrocyanation of a Sterically Hindered Aldehyde. Optimization of a Chemoenzymatic Procedure for (*R*)-2-Chloromandelic Acid. *Org. Process Res. Dev.* **2003**, *7*, 828–831. [CrossRef]

25. Veum, L.; Pereira, S.R.M.; Van Der Waal, J.C.; Hanefeld, U. Catalytic Hydrogenation of Cyanohydrin Esters as a Novel Approach to N-Acylated β-Amino Alcohols-Reaction Optimisation by a Design of Experiment Approach. *Eur. J. Org. Chem.* **2006**, *2006*, 1664–1671. [CrossRef]

26. Cheng, M.J.; Lee, K.H.; Tsai, I.L.; Chen, I.S. Two New Sesquiterpenoids and Anti-HIV Principles from the Root Bark of *Zanthoxylum ailanthoides*. *Bioorg. Med. Chem.* **2005**, *13*, 5915–5920. [CrossRef] [PubMed]

27. Shoeb, A.; Kapil, R.S.; Popli, S.P. Coumarins and Alkaloids of *Aegle marmelos*. *Phytochemistry* **1973**, *12*, 2071–2072. [CrossRef]

28. Semba, H.; Dobashi, Y.; Matsui, T. Expression of Hydroxynitrile Lyase from *Manihot esculenta* in Yeast and Its Application in (*S*)-Mandelonitrile Production Using an Immobilized Enzyme Reactor. *Biosci. Biotechnol. Biochem.* **2008**, *72*, 1457–1463. [CrossRef]

29. Müller, J.; Sowa, M.A.; Fredrich, B.; Brundiek, H.; Bornscheuer, U.T. Enhancing the Acyltransferase Activity of *Candida antarctica* Lipase A by Rational Design. *ChemBioChem* **2015**, *16*, 1791–1796. [CrossRef]

30. Phillips, R.S. Temperature Modulation of the Stereochemistry of Enzymatic Catalysis: Prospects for Exploitation. *Trends Biotechnol.* **1996**, *14*, 13–16. [CrossRef]

31. Dadashipour, M.; Asano, Y. Hydroxynitrile Lyases: Insights into Biochemistry, Discovery and Engineering. *ACS Catal.* **2011**, *1*, 1121–1149. [CrossRef]

32. Willeman, W.F.; Straathof, A.J.J.; Heijnen, J.J. Reaction Temperature Optimization Procedure for the Synthesis of (*R*)-Mandelonitrile by *Prunus amygdalus* Hydroxynitrile Lyase Using a Process Model Approach. *Enzyme Microb. Technol.* **2002**, *30*, 200–208. [CrossRef]

33. Ueatrongchit, T.; Tamura, K.; Ohmiya, T.; H-Kittikun, A.; Asano, Y. Hydroxynitrile Lyase from *Passiflora edulis*: Purification, Characteristics and Application in Asymmetric Synthesis of (*R*)-Mandelonitrile. *Enzyme Microb. Technol.* **2010**, *46*, 456–465. [CrossRef] [PubMed]

34. Willeman, W.F.; Hanefeld, U.; Straathof, A.J.J.; Heijnen, J.J. Estimation of Kinetic Parameters by Progress Curve Analysis for the Synthesis of (*R*)-Mandelonitrile by *Prunus amygdalus* Hydroxynitrile Lyase. *Enzyme Microb. Technol.* **2000**, *27*, 423–433. [CrossRef]

35. O'Neil, M.J.; Patricia, E.; Heckelman, A.S.; Budavari, S. *The Merck Index-An Encyclopedia of Chemicals Drugs and Biologicals*; Merck and Co.: Whitehouse Station, NJ, USA, 2006; p. 830.

36. Shakeri, M.; Engström, K.; Sandström, A.G.; Bäckvall, J.E. Highly Enantioselective Resolution of β-Amino Esters by *Candida antarctica* Lipase A Immobilized in Mesocellular Foam: Application to Dynamic Kinetic Resolution. *ChemCatChem* **2010**, *2*, 534–538. [CrossRef]

37. Ding, W.; Li, M.; Dai, R.; Deng, Y. Lipase-Catalyzed Synthesis of the Chiral Tetrahydroisoquinoline (*R*)-Salsolinol. *Tetrahedron Asymmetry* **2012**, *23*, 1376–1379. [CrossRef]

38. Zamost, B.L.; Nielsen, H.K.; Starnes, R.L. Thermostable Enzymes for Industrial Applications. *J. Ind. Microbiol.* **1991**, *8*, 71–81. [CrossRef]

39. Adlercreutz, P. Immobilisation and Application of Lipases in Organic Media. *Chem. Soc. Rev.* **2013**, *42*, 6406–6436. [CrossRef] [PubMed]

40. Bastida, A.; Sabuquillo, P.; Armisen, P.; Fernández-Lafuente, R.; Huguet, J.; Guisán, J.M. A Single Step Purification, Immobilization, and Hyperactivation of Lipases via Interfacial Adsorption on Strongly Hydrophobic Supports. *Biotechnol. Bioeng.* **1998**, *58*, 486–493. [CrossRef]

41. Hanefeld, U. Immobilisation of Hydroxynitrile Lyases. *Chem. Soc. Rev.* **2013**, *42*, 6308–6321. [CrossRef]

42. Cabirol, F.L.; Hanefeld, U.; Sheldon, R.A. Immobilized Hydroxynitrile Lyases for Enantioselective Synthesis of Cyanohydrins: Sol-Gels and Cross-Linked Enzyme Aggregates. *Adv. Synth. Catal.* **2006**, *348*, 1645–1654. [CrossRef]

43. Valivety, R.H.J.; Halling, P.; Macrae, A.R. Reaction Rate with Suspended Lipase Catalyst Shows Similar Dependence on Water Activity in Different Organic Solvents. *Biochim. Biophys. Acta (BBA)/Protein Struct. Mol.* **1992**, *1118*, 218–222. [CrossRef]

44. Schmitke, J.L.; Wescott, C.R.; Klibanov, A.M. The Mechanistic Dissection of the Plunge in Enzymatic Activity upon Transition from Water to Anhydrous Solvents. *J. Am. Chem. Soc.* **1996**, *118*, 3360–3365. [CrossRef]

45. Wehtje, E.; Costes, D.; Adlercreutz, P. Enantioselectivity of Lipases: Effects of Water Activity. *J. Mol. Catal. B Enzym.* **1997**, *3*, 221–230. [CrossRef]

46. Pepin, P.; Lortie, R. Influence of Water Activity on the Enantioselective Esterification of (*R,S*)-Ibuprofen by *Candida antarctica* Lipase B in Solventless Media. *Biotechnol. Bioeng.* **1999**, *63*, 502–505. [CrossRef]

47. Bovara, R.; Carrea, G.; Ottolina, G.; Riva, S. Water Activity Does Not Influence the Enantioselectivity. *Biotechnol. Lett.* **1993**, *15*, 169–174. [CrossRef]

48. Carrea, G.; Riva, S. Properties and Synthetic Applications of Enzymes in Organic Solvents. *Angew. Chem. Int. Ed. Engl.* **2000**, *39*, 2226–2254. [CrossRef]

49. Lauble, H.; Miehlich, B.; Förster, S.; Wajant, H.; Effenberger, F. Crystal Structure of Hydroxynitrile Lyase from *Sorghum bicolor* in Complex with the Inhibitor Benzoic Acid: A Novel Cyanogenic Enzyme. *Biochemistry* **2002**, *41*, 12043–12050. [CrossRef]
50. Xu, L.L.; Singh, B.K.; Conn, E.E. Purification and Characterization of Mandelonitrile Lyase from *Prunus lyonii*. *Arch. Biochem. Biophys.* **1986**, *250*, 322–328. [CrossRef]
51. Jaenicke, L.; Preun, J. Chemical Modification of Hydroxynitrile Lyase by Selective Reaction of an Essential Cysteine-SH Group with α,β-unsaturated Propiophenones as Pseudo-substrates. *Eur. J. Biochem.* **1984**, *138*, 319–325. [CrossRef]
52. Hanefeld, U.; Straathof, A.J.J.; Heijnen, J.J. Study of the (S)-Hydroxynitrile Lyase from *Hevea brasiliensis*: Mechanistic Implications. *Biochim. Biophys. Acta—Protein Struct. Mol. Enzymol.* **1999**, *1432*, 185–193. [CrossRef]
53. Lowe, M.E. Assays for Pancreatic Triglyceride Lipase and Colipase. In *Lipase and Phospholipase Protocols. Methods in Molecular Biology*; Doolittle, M., Reue, K., Eds.; Humana Press: New York, NY, USA, 1999; Volume 109, pp. 59–70.
54. Schallmey, M.; Jekel, P.; Tang, L.; Elenkov, M.M.; Höffken, H.W.; Hauer, B.; Janssen, D.B. A single point mutation enhances hydroxynitrile synthesis by halohydrin dehalogenase. *Enzyme Microb. Technol.* **2015**, *70*, 50–57. [CrossRef]
55. Chen, C.; Fujimoto, Y.; Girdaukas, G.; Sih, C.J. Quantitative Analyses of Biochemical Kinetic Resolutions of Enantiomers. *J. Am. Chem. Soc.* **1982**, *104*, 7294–7299. [CrossRef]
56. Sakuth, M.; Mensing, T.; Schuler, J.; Heitmann, W.; Strehlke, G.; Mayer, D. Ethers, Aliphatic. In *Ullmann's Encyclopedia of Industrial Chemistry*, 6th ed.; Elvers, B., Ed.; Wiley-VCH Verlag GmbH & Co. KGaA: Weinheim, Germany, 2012; p. 440.
57. Halling, P.J. Salt Hydrates for Water Activity Control with Biocatalysts in Organic Media. *Biotechnol. Tech.* **1992**, *6*, 271–276. [CrossRef]

Article

Production and Surfactant Properties of *Tert*-Butyl α-ᴅ-Glucopyranosides Catalyzed by Cyclodextrin Glucanotransferase

Humberto Garcia-Arellano [1,2], Jose L. Gonzalez-Alfonso [1], Claudia Ubilla [1,3], Francesc Comelles [4], Miguel Alcalde [1], Manuel Bernabé [5,†], José-Luis Parra [4], Antonio O. Ballesteros [1] and Francisco J. Plou [1,*]

[1] Instituto de Catálisis y Petroleoquímica, CSIC, 28049 Madrid, Spain
[2] Departamento de Ciencias Ambientales, División de Ciencias Biológicas y de la Salud, Universidad Autónoma Metropolitana, Unidad Lerma, Av. de las Garzas 10, Lerma de Villada, Estado de México 52006, Mexico
[3] Escuela de Ingeniería Bioquímica, Pontificia Universidad Católica de Valparaíso, Valparaíso 2147, Chile
[4] Instituto de Investigaciones Químicas y Ambientales, CSIC, 08034 Barcelona, Spain
[5] Instituto de Química Orgánica, CSIC, 28006 Madrid, Spain
* Correspondence: fplou@icp.csic.es; Tel.: +34-915854869
† Dedicated to the memory of Prof. Manuel Bernabé, deceased on April 2018.

Received: 28 May 2019; Accepted: 26 June 2019; Published: 29 June 2019

Abstract: While testing the ability of cyclodextrin glucanotransferases (CGTases) to glucosylate a series of flavonoids in the presence of organic cosolvents, we found out that this enzyme was able to glycosylate a tertiary alcohol (*tert*-butyl alcohol). In particular, CGTases from *Thermoanaerobacter* sp. and *Thermoanaerobacterium thermosulfurigenes* EM1 gave rise to the appearance of at least two glycosylation products, which were characterized by mass spectrometry (MS) and nuclear magnetic resonance (NMR) as *tert*-butyl-α-D-glucoside (major product) and *tert*-butyl-α-D-maltoside (minor product). Using partially hydrolyzed starch as glucose donor, the yield of transglucosylation was approximately 44% (13 g/L of *tert*-butyl-α-D-glucoside and 4 g/L of *tert*-butyl-α-D-maltoside). The synthesized *tert*-butyl-α-D-glucoside exhibited the typical surfactant behavior (critical micellar concentration, 4.0–4.5 mM) and its properties compared well with those of the related octyl-α-D-glucoside. To the best of our knowledge, this is the first description of an enzymatic α-glucosylation of a tertiary alcohol.

Keywords: biocatalysis; glycosidases; transglycosylation; cyclodextrin glycosyltransferases; alkyl glucosides; biosurfactants

1. Introduction

Cyclodextrin glucanotransferases—also known as cyclodextrin glycosyltransferases (CGTases, EC 2.4.1.19)—are extracellular enzymes included in the so-called α-amylase family GH13 [1] and are able to convert starch and related maltodextrin substrates into nonreducing, cyclic glucooligosaccharides termed cyclodextrins (CDs) [2]. CDs are formed by an intramolecular transglycosylation reaction in which 6, 7, or 8 glucosyl residues are linked by α(1→4) glycosidic bonds, giving rise to the formation of α, β, or γ-CDs, respectively [3]. Besides cyclization, CGTases also catalyze three additional reactions: Coupling, i.e., the aperture of the CD ring and subsequent transfer of glucose residues to acceptors; disproportionation, i.e., the transfer reaction between two linear dextrins to form maltooligosaccharides of different sizes; and hydrolysis, in which the acceptor is a H_2O molecule [4]. It is worth noting that CGTases display a weak hydrolyzing activity (1.5–3.0 U/mg protein) [5]. The main reaction catalyzed by these enzymes (in terms of specific activity) is the intermolecular transglycosylation (coupling and disproportionation, employing CDs or maltodextrins as glucosyl donors, respectively) [6,7]. In the presence of certain carbohydrates such as sucrose, CGTase also transfers glucosyl moieties yielding the so-called acceptor products [8].

The enzyme CGTase has proved an exceptional capability to glucosylate compounds of different nature employing starch, maltodextrins, or cyclodextrins as glucosyl donors [9,10]. Apart from monosaccharides and disaccharides [11,12], other compounds such as flavonoids [13,14], vitamins [15], sugar alcohols [16], sweet glycosides [17], and polyols [18] have been successfully used as acceptor molecules for the intermolecular transglycosylation. Unfortunately, one of the main drawbacks of CGTases is that their product selectivity is not very high, because the enzyme displays its four activities simultaneously [19]. Furthermore, the formation of a homologous series of polyglucosylated products is normally described with this enzyme [20–22]. As a result, a low yield of the desired glycosylated product is typically obtained [23]. Several strategies have been assessed to improve the transglycosylation activity of CGTases, including protein engineering [24–26], chemical modification [7], immobilization [27,28], and the addition of cosolvents and additives [29–31]. Bacterial CGTases are produced mainly by the genus *Bacillus* [32], although *Micrococcus* and *Klebsiella* species had also been reported as producers.

For many of these acceptor reactions catalyzed by CGTases and other glycosidic enzymes, a cosolvent is needed to increase the solubility of the acceptor in the reaction medium. A requirement of the solvent is that it cannot act as acceptor itself, because this could cause a reduction of the yield of the transglycosylation and lead to the appearance of undesired side products. For this reason, primary and secondary alcohols are barely used in these transglycosylations. In fact, the synthesis of alkyl glycosides has been widely reported using the reverse hydrolysis reaction catalyzed by glycosidases employing mainly primary and secondary alcohols as acceptors [33–35]. Typical cosolvents for transglycosylation reactions include acetonitrile, DMSO, ethers, tertiary alcohols, etc. [22,23,36], and more recently, biomass-derived solvents and ionic liquids [37,38].

In this work, while testing the ability of CGTases to glucosylate a series of flavonoids in presence of organic cosolvents, we found out that this enzyme was able to glycosylate a tertiary alcohol (*tert*-butyl alcohol). We report herein the enzymatic synthesis of two novel alkyl glycoside derivatives, namely *tert*-butyl-α-D-glucoside and *tert*-butyl α-D-maltoside. Alkyl glucosides (AGs) are non-ionic surface-active agents with extensive applications in food, cosmetic, and detergent industries [39,40], in part due to their antimicrobial activity, biodegradability, and low toxicity. The surfactant properties of the major synthesized product were compared with those of related alkyl glycosides.

2. Results and Discussion

2.1. Glucosylation of Tert-Butyl Alcohol by CGTases and Characterization of Products

In our laboratory we are studying the application of different CGTases to synthesize glucosylated derivatives of polyphenols with bioactive properties [41]. Many of these flavonoids are scarcely soluble in water, which implies the need of using a miscible cosolvent to increase the solubility of the acceptor. While screening a series of cosolvents (at 30% v/v) for the glycosylation of polyphenols, we observed in the HPLC chromatogram, when using *tert*-butyl alcohol, the appearance of several peaks with retention times lower than the corresponding to glucose (Figure 1). Such peaks were also present in the control reaction in absence of the acceptor, but not in the control experiments lacking enzyme, sugar donor, or *tert*-butyl alcohol. Thus, it seemed that CGTase was able to glycosylate the tertiary alcohol *tert*-butyl alcohol.

We screened four CGTases from different bacterial species (Table 1) for their transglycosylation efficiency towards *tert*-butyl alcohol at 60 °C. First, the transglycosylation activity of these enzymes was determined with a test based on the use of p-nitrophenyl-α-D-maltoheptaoside-4,6-O-ethylidene (EPS) as glucosyl donor and maltose as acceptor [12]. For commercial CGTases, the preparations were partially purified by a PD-10 desalting column (GE Healthcare, Chicago, IL, USA) to eliminate low-molecular-weight contaminants that could interfere with our reaction. The transglycosylation experiments towards *tert*-butyl alcohol were carried out employing the same amount of EPS enzyme units.

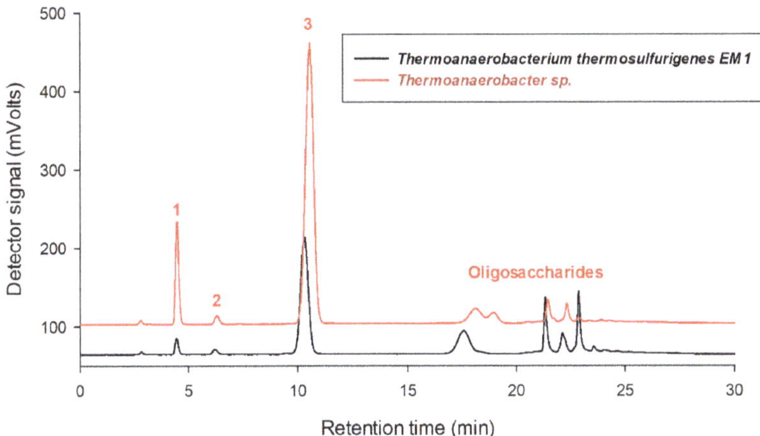

Figure 1. HPLC chromatogram of the reaction mix after 24 h displaying the acceptor products of *tert*-butyl alcohol synthesized with CGTases from *Thermoanaerobacter* sp. and *Thermoanaerobacterium thermosulfurigenes* EM1. Reaction conditions: Soluble starch (30 g/L), 30% *tert*-butyl alcohol, 10 mM sodium citrate buffer (pH 5.5), CGTase (0.4 U/mL, EPS method), and 60 °C. (1) Main glucosylation product; (2) minor glucosylation product; and (3) glucose.

Table 1. Cyclodextrin glucanotransferases (CGTases) screened for the glucosylation of *tert*-butyl alcohol.

Source	Protein (mg/mL)	Transglycosylation Activity (U/mL) [a]
Thermoanaerobacter sp.	2.88	37.1
Thermoanaerobacterium thermosulfurigenes EM1	0.77	43.1
Geobacillus sp.	2.63	66.4
Bacillus circulans 251	2.79	152.7

[a] Measured by the EPS method.

The TLC screening showed that the only enzymes that led to a noticeable production of glucosylated products, under the tested conditions, were the CGTases from *Thermoanaerobacter* sp. (Toruzyme 3.0L) and *Thermoanaerobacterium thermosulfurigenes* EM1. No appreciable formation of the new acceptor derivatives was observed with the rest of the CGTases. Moreover, in the case of *Bacillus circulans* and *Geobacillus* sp. CGTases, *tert*-butyl alcohol seemed to produce an inhibitory effect on enzyme activity. Figure 1 shows the HPLC chromatograms obtained with the CGTases from *T. thermosulfurigenes* EM1 and *Thermoanaerobacter* sp. after 24 h of reaction.

Several carbohydrates (maltose, α-cyclodextrin, β-cyclodextrin, and partially hydrolyzed starch) were compared as sugar donors in the transglycosylation reaction towards *tert*-butyl alcohol with *Thermoanaerobacter* sp. CGTase. All the substrates tested yielded the expected transglycosylation products. However, partially hydrolyzed starch was selected as the best glucosyl donor for further experiments due to its availability and lower cost.

With the aim to clearly identify the nature of the new synthesized products, the two main peaks in the chromatograms of Figure 1 were isolated by semipreparative HPLC. Both products were purified at a high degree (>98%). The major product (**1**) was a white solid whereas the minor derivative (**2**) showed an oily appearance. The molecular weight was established by MS-QTOF using ionization by electrospray (see Supplementary Materials, Figures S1 and S2). For compound **1**, the main peak in the MS spectrum (positive mode) was at *m/z* 259.11 that corresponded to the M+[Na]$^+$ ion of the *tert*-butyl glucoside. For compound **2**, the major signal of the MS spectrum in positive mode was at *m/z* 421.17 that fits with the M+[Na]$^+$ ion of the *tert*-butyl maltoside. The identity of the products was clearly established by ^{13}C and ^{1}H NMR analysis. NMR data confirmed the α-configuration of the glucosyl moieties. The structure of the synthesized compounds is represented in Figure 2. The major compound was *tert*-butyl-O-α-D-glucopyranoside and the minor product the corresponding α-D-maltoside. A particular feature of CGTase is the formation of several polyglucosylated products (the so-called homologous series) with different acceptors [8,42].

***tert*-butyl-O-α-D-glucopyranoside.** 1H-NMR (δ): 5.07 (d, 1H, H-1, $J_{1,2}$ = 3.9 Hz); 3.78-3.66 (m, 3H, H-5 + H-6a + H-6b); 3.61 (t, 1H, H-3, $J_{2,3}$ = 9.4, $J_{3,4}$ = 9.2 Hz); 3.31 (dd, 1H, H-2); 3.28 (t, 1H, H-4, $J_{4,5}$ = 9.3 Hz); 1.30 (s, 9H, t-butyl). 13C-NMR (δ): 94.6 (C-1); 75.2 (C-3); 73.5 (C-5); 73.1 (C-2); 72.0 (C-4); 62.7 (C-6); 28.9 (*tert*-butyl).

***tert*-butyl-O-α-D-maltoside.** 1H-NMR (δ): 5.13 (d, 1H, H-1′, $J_{1',2'}$ = 3.9 Hz); 5.06 (d, 1H, H-1, $J_{1,2}$ = 3.8 Hz); 3.89-3.64 (m, 6H, H-5 + H-6a + H-6b + H-5′ + H-6′a + H′-6b); 3.82 (dd, 1H, H-3, $J_{2,3}$ = 9.6, $J_{3,4}$ = 9.0 Hz); 3.62 (t, 1H, H-3′, $J_{2',3'}$ = 9.9, $J_{3',4'}$ = 9.0 Hz); 3.50 (t, 1H, H-4, $J_{4,5}$ = 9.8 Hz); 3.43 (dd, 1H, H-2′): 3.37 (dd, 1H, H-2); 3.26 (t, 1H, H-4′, $J_{4',5'}$ = 9.3 Hz); 1.27 (s, 9H, t-butyl). 13C-NMR (δ): 102.9 (C-1′); 94.5 (C-1); 81.9 (C-4); 75.1 + 75.0 (C-3′ + C-5′); 74.7 (C-2′); 74.3 (C-3); 73.1 (C-2); 71.8 (C-5); 71.5 (C-4′); 62.8 (C-6); 62.1 (C-6′); 28.9 (*tert*-butyl).

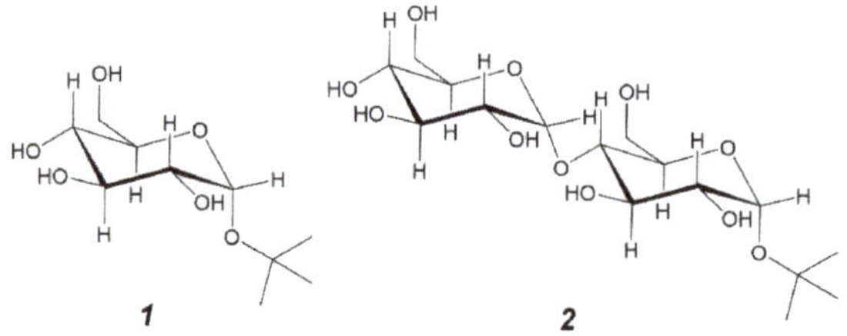

Figure 2. Chemical structure of the synthesized glycosides: (**1**) *tert*-butyl-O-α-D-glucopyranoside and (**2**) *tert*-butyl-O-α-D-maltoside.

It is remarkable that CGTases can use a sterically hindered alcohol as acceptor. Alcohols have been frequently used as additives in the reaction media to improve or direct the synthesis of specific cyclodextrins by CGTases. However, to the extent of our knowledge, no other reports describe the synthesis of alkyl glucosides catalyzed by CGTases. Svensson et al. synthesized dodecyl-β-maltooctaoside with CGTase from *B. macerans* by an indirect method based on the lengthening of dodecyl-β-maltoside using α-cyclodextrin as glucosyl donor [43]. In this context, CGTases are able to transglycosylate polyhydroxylated compounds such as trimethylolpropane or glycerol [44].

The synthesis of alkyl glycosides has been manly achieved using the reverse hydrolysis reaction catalyzed by glycosidases [45,46]. Glycosidases show remarkable chemoselectivity for primary and secondary alcohols as well as phenols, but tertiary alcohols are not easily recognized as substrates. For the glycosylation of phenols, CGTase is one of the best options because its hydrolytic activity is very low. Since phenol is a good leaving group, glycosidases typically fast hydrolyze the synthesized phenyl glycosides, thus lowering the yield.

In general, secondary alcohols are glycosylated more slowly than primary ones by a factor of 3–5 [33]. Tertiary alcohols were considered unreactive in such enzymatic reactions until 1996, when Fischer et al. demonstrated that a tertiary alcohol (2-methyl-2-butanol) could act as nucleophile in a reaction catalyzed by the β-glucosidase from *Pyrococcus furiosus*, using cellobiose as glucosyl donor [47]. Later, Svasti et al. described the synthesis of tertiary β-glucosides catalyzed by a β-glucosidase from cassava [34]. However, this enzyme requiered activated *p*-nitrophenyl (pNP) glycosides as sugar donors and was not efficient with mono- and disaccharides. In addition, the synthesized β-glucoside was fast hydrolyzed as a consequence of the hydrolytic activity of the enzyme. Jiang et al. were able to synthesize *tert*-butyl β-D-xyloside and xylobioside with xylan as sugar donor, in the presence of 20% *tert*-butyl alcohol, employing a xylanase from *Thermotoga maritima* [48]. Kongsaree et al. reported in 2010 that linamarase (a cyanogenic β-glucosidase) also catalyzed transglucosylation reactions with tertiary alcohols as acceptors and pNP-glycosides as donors [49].

Regarding the formation of alkyl glycosides of tertiary alcohols with α-configuration, Simerska et al. were the first in reporting the enzymatic α-glycosylation of two sterically hindered alcohols, namely *tert*-butyl and *tert*-amyl alcohol; the enzyme was a α-galactosidase from *Talaromyces flavus*, with pNP α-galactoside as donor, yielding the corresponding α-D-galactopyranosides [50]. To our knowledge, our work is the first report of an enzymatic α-glucosylation of a tertiary alcohol. The chemical synthesis of *tert*-butyl α-glucosides is quite complex and usually gives rise to a mixture of anomers [51].

2.2. Progress of Tert-Butyl Alcohol Glucosylation by Thermoanaerobacter sp. CGTase

In order to dismiss that none of the contaminant enzymes in the Toruzyme preparation was responsible of *tert*-butyl alcohol glycosylation, the CGTase from *Thermoanaerobacter* sp. was purified to homogeneity by sequential chromatography steps on DEAE-Sepharose and α-cyclodextrin-activated Superose. We observed that the introduction of an ion-exchange chromatography step before the commonly used affinity step rendered a pure enzyme as confirmed by SDS-PAGE (Figure 3).

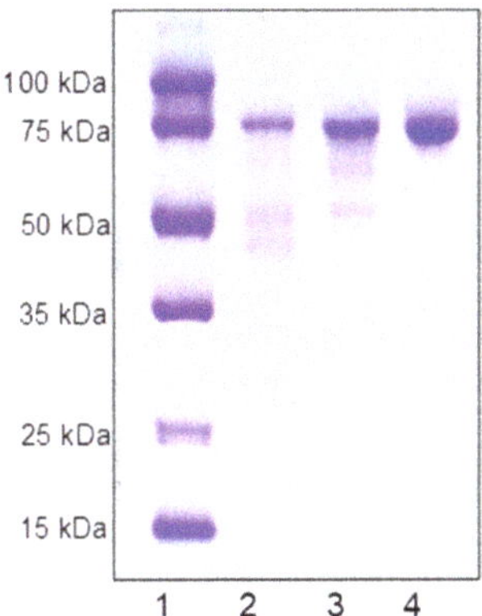

Figure 3. Purification of CGTase from *Thermoanaerobacter* sp. Lane 1, molecular mass markers; lane 2, crude extract; lane 3, DEAE chromatography step; and lane 4, affinity chromatography step.

We evaluated the production of *tert*-butyl-glucosides with the purified enzyme in a batch reactor employing 30% *tert*-butyl alcohol. Although *Thermoanaerobacter* sp. CGTase has an optimal temperature around 90 °C [52], the reaction was carried out at 60 °C to avoid starch browning and evaporation over time. Figure 4 depicts the progress of the reaction. Approximately 13 g/L of *tert*-butyl glucoside were produced after 200 h of reaction. In the case of the maltoside derivative, a plateau after 60 h was reached yielding only 4 g/L. Considering that the initial concentration of soluble starch was 30 g/L, and that the sugar donor is the limiting reagent in the reaction, the yield of the *tert*-butyl glycosides was approximately 44%.

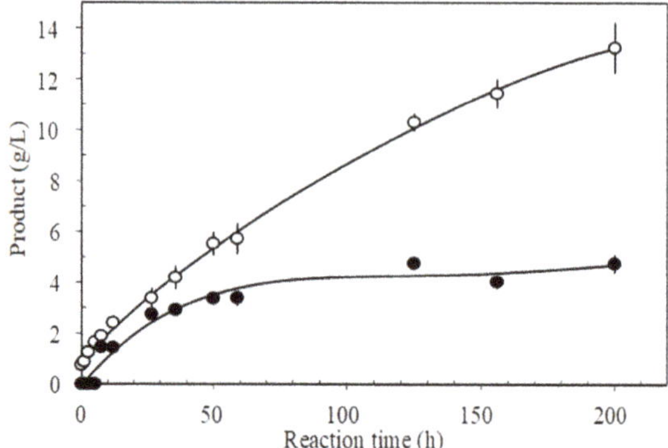

Figure 4. Progress of *tert*-butyl alcohol glucosylation by CGTase from *Thermoanaerobacter* sp. Open circles, α-glucoside derivative; filled circles, α-maltoside derivative. Reaction conditions: Soluble starch (30 g/L), 30% *tert*-butyl alcohol, 10 mM citrate buffer (pH 5.5), 0.4 U/mL CGTase (transglycosylation activity, EPS method), and 60 °C.

2.3. Surfactant Properties of Tert-Butyl Glucoside

Anomerically pure alkyl glucosides (AGs) are very useful in pharmaceutical and biomedical applications [53] due to their compatibility with biological systems that arise from the presence of the sugar head groups attached to the molecule. AGs do not denature enzymes and proteins, and in consequence are commonly used in protein extraction from cellular membranes [54]. AGs, unlike other sugar-based surfactants, are quite stable under alkaline conditions. In previous work, while studying several resveratrol glucosides, we demonstrated that the α-configuration gave rise to a superior surfactant performance when compared with β-configuration [9].

The surfactant properties of the monoglucosylated derivative were measured and compared with the reference compound octyl-α-D-glucoside (Figure 5). Our results suggest a similar surfactant behavior for both compounds. The synthesized *tert*-butyl glucoside behaves as a typical surfactant, defined by a linear descent of the surface tension vs. the logarithmic concentration of the surfactant, reaching a well-defined critical micellar concentration (CMC). CMC values of the *tert*-butyl and octyl α-derivatives were closely similar. It is worth noting that the relatively small *tert*-butyl group seems to be equivalent to the bulky octyl group for micelle formation.

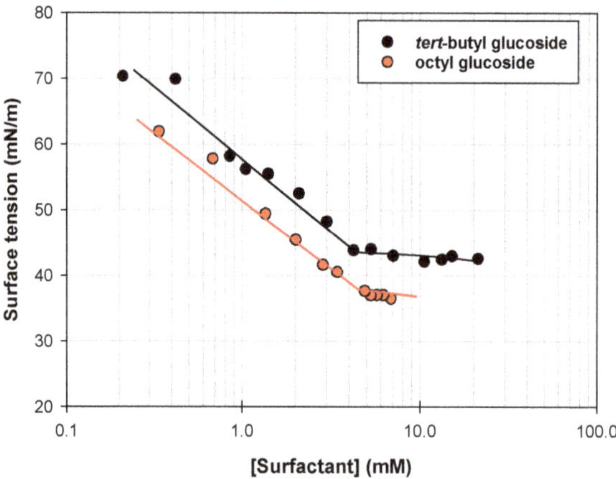

Figure 5. Variation of surface tension vs. concentration for *tert*-butyl and octyl glucosides.

The main surfactant parameters of both compounds are summarized in Table 2. The surface tension at CMC is slightly higher for the *tert*-butyl glucoside with respect to the octyl derivative. As expected, the average area occupied per each molecule of adsorbed surfactant at the saturated water-air interface (*A*) was higher for the *tert*-butyl glucoside compared with the octyl glucoside. The parameter *p*C20, which is directly related to the efficiency of a surfactant, was only 10% higher for octyl-α-glucoside compared with the *tert*-butyl derivative.

Table 2. Surfactant properties of *tert*-butyl-O-α-D-glucoside and octyl-O-α-D-glucoside.

Property	*Tert*-butyl-α-glucoside	Octyl-α-glucoside
CMC (mM)	4.0–4.5	4.8–5.0
Surface tension at CMC (mN/m)	43.9	37.0
*p*C20 [a]	2.72	3.00
A (Å²) [b]	66.5	45.1

[a] Minus logarithm of the concentration required to diminish the surface tension of water by 20 mN/m. [b] Area occupied per molecule adsorbed at the saturated interface.

3. Materials and Methods

3.1. Enzymes and Reagents

CGTase from *Thermoanaerobacter* sp. (Toruzyme 3.0L) was kindly provided by Novozymes A/S (Bagsværd, Denmark). CGTase from *Geobacillus* sp. (CGT-SL) was from Amano Enzyme Inc. (Nagoya, Japan). Both CGTases were partially purified using a PD-10 desalting column (GE Healthcare). CGTases from *T. thermosulfurigenes* EM1 and *B. circulans* strain 251 were kindly provided in purified form by L. Dijkhuizen (Groningen University, The Netherlands). The α-glucosidase (EC 3.2.1.20) from *Saccharomyces cerevisiae* was from Boehringer Mannheim. β-Cyclodextrin was purchased from Sigma. *p*-Nitrophenyl-α-D-maltoheptaoside-4,6-O-ethylidene (EPS) was from Boehringer Mannheim. Starch from potato Paselli SA2 (partially hydrolyzed with a mean degree of polymerization of 50) was kindly provided by Avebe (Foxhol, The Netherlands). *Tert*-butyl alcohol was from Fluka. All other reagents and solvents were of the highest purity grade available.

3.2. Activity Assay

The transglycosylation (disproportionation) activity of CGTase was measured on the basis of the method described by Nakamura et al. [55], adapted to 96-well plates. EPS (*p*-nitrophenyl-α-D-maltoheptaoside-4-6-O-ethylidene) was as the glucosyl donor and maltose the acceptor. In this assay EPS is first cleaved and the maltose molecule is then linked to the free reducing end. Thereafter, the *p*-nitrophenol is released from the reaction product by the action of α-glucosidase. In particular, 20 µL of a maltose solution (50 mM) were mixed with 10 µL of an EPS solution (30 mM) and 65 µL of 0.2 M phosphate buffer (pH 7.0). The reaction, performed by triplicate, started when 5 µL of enzymatic solution was added. The 96-well plate was incubated at 60 °C for 20 min. The transglycosylation was stopped with 10 µL of 1 M HCl and the mixture was incubated at 60 °C for 10 min. Thereafter it was neutralized with 10 µL of 1 M NaOH. The α-glucosidase (EC 3.2.1.20) from *S. cerevisiae* was then added (0.005 U) and the mixture was kept at 25 °C for 1 h. Finally, Na_2CO_3 (1 M) was added to a final volume of 200 µL and the absorbance was measured at 401 nm. A calibration curve with *p*-nitrophenol was made from a stock solution of 0.74 mg/mL. Different volumes between 0–10 µL were taken by triplicate and conveniently diluted with 0.2 M phosphate buffer to a total volume of 100 µL. One unit of activity (U) was established as that corresponding to the release of 1 µmol of *p*-nitrophenol per min under these conditions.

3.3. Enzymatic Synthesis of Tert-Butyl Glucosides

Typical reaction mixtures contained 30% v/v *tert*-butyl alcohol, a 3% w/v sugar donor (Paselli SA2 starch, maltose, or cyclodextrins) and 0.4 U/mL CGTase (transglycosylation activity, chapter 3.2), in 10 mM sodium citrate buffer (pH 5.5). Reaction mixtures were incubated at 250 rpm in an orbital incubator at 60 °C. At regular time intervals, aliquots were removed and analyzed by TLC and HPLC. Controls without enzyme, *tert*-butyl alcohol, and/or sugar donor were also performed to rule out the presence of non-enzymatic products.

3.4. TLC and HPLC Analyses

Reaction samples were filtered on UltraFree centrifugal filters (0.45 μm, Millipore, Burlington, MA, USA) and analyzed in silica gel plates (10 × 5 cm, Merck) developed with ethyl acetate/methanol/water (77/15/8, v/v/v). Spots containing the residual carbohydrates and products were visualized after staining with a p-methoxybenzaldehyde solution (ethanol/sulphuric acid/p-methoxybenzaldehyde, 19/1/1, v/v/v) and heated at 90 °C for 15 min. The reaction progress was followed by normal-phase high-performance liquid chromatography (HPLC) employing a quaternary pump (Delta 600, Waters, Milford, MA, USA) and a Luna NH2 column (250 mm × 4.6 mm, 5 μm, Phenomenex). The starting mobile phase was acetonitrile:water 78:22 (v/v) and the gradient is outlined in Table 3. The solvents were conditioned with helium and the flow rate was 1.0 mL/min. The temperature of the column was kept constant at 30 °C. An evaporative light-scattering detector (model PL-ELS 1000, Polymer Laboratories, Salop, UK) was used and fixed to a nebulization and evaporation temperatures of 80 °C and 90 °C, respectively. The data obtained were analyzed using the Millennium Software, employing purified compounds as external standards for calibration.

Table 3. HPLC gradient profile.

Time	Acetonitrile	Water
0–12 min	78%	22%
12–15 min	78% → 50%	22% → 50%
15–20 min	50%	50%
20–21 min	50% → 78%	50% → 22%
21–30 min	78%	22%

3.5. Purification of Tert-Butyl Glucosides

The reaction mixture (100 mL) contained 30% v/v *tert*-butyl alcohol, 3% w/v Paselli SA2 (sugar donor), and 0.4 U/mL CGTase (transglycosylation activity) in 10 mM sodium citrate buffer (pH 5.5). The mixture was incubated at 250 rpm in an orbital shaker and 60 °C. After 200 h the reaction was stopped, and products were purified as follows. Reaction medium was filtered (0.45 μm, Millipore) and water and residual *tert*-butyl alcohol were evaporated by rotary evaporation at 60 °C. The dried products were then dissolved in a mixture of acetonitrile:water (88:12 v/v) and purified in a semipreparative HPLC pump (Delta 600, Waters) coupled to a Kromasil-NH2 column (250 mm × 10 mm, Analisis Vinicos, Spain) and to a refraction-index detector (model 9040, Varian, Palo Alto, CA, USA). Compounds were eluted in an isocratic regime using acetonitrile:water (88:12 v/v) as mobile phase at a flow rate of 8.0 mL/min. Fractions containing the desired products were collected and pooled. The mobile phase was evaporated by rotary evaporation at 60 °C and the dried products were stored at −20 °C. The purity of the isolated compounds was assessed by analytical HPLC. The equivalent isolated yields were approximately 1.1 g of *tert*-butyl-glucoside and 0.2 g of *tert*-butyl-maltoside.

3.6. Mass Spectrometry (MS)

The MS analysis of purified glucosides was carried out with a mass spectrometer equipped with hybrid QTOF analyzer (QSTAR, Pulsar i, AB Sciex, Madrid, Spain). Samples were analyzed by direct infusion using electrospray ionization in positive reflector mode. Methanol containing 1% of NH_4OH was employed as ionizing phase.

3.7. Nuclear Magnetic Resonance (NMR)

NMR spectra were determined on a Varian Unit (1H-NMR, 500 MHz; 13C-NMR, 125 MHz) spectrometer, equipped with a gradient unit and a reverse probe. Samples (6–10 mg) were dissolved in 0.6 mL of semiheavy water (HDO and the spectra were obtained at 40 °C. Proton chemical shifts were referred to residual HDO (4.61 ppm). Carbon chemical shifts were referred to external acetone (31.07 ppm). 2D-homo- (DQCOSY, TOCSY (HOHAHA), NOESY) and hetero- (HMQC and HMBC) NMR experiments were carried out by using the standard software from Varian.

3.8. Enzyme Purification

Thermoanaerobacter sp. CGTase was purified from the commercial extract (Toruzyme 3.0L, Novozymes A/S, Bagsværd, Denmark)) through successive steps of ion exchange and affinity chromatography on a FPLC LCC-500 equipment (Pharmacia, Sweden). The commercial extract was dialyzed against 20 mM Tris-HCl buffer (pH 7.5) and loaded onto a DEAE-Sepharose column equilibrated with the same buffer. Proteins were eluted using a linear salt gradient from 0 to 1 M of NaCl in the Tris-HCl buffer. The fractions containing CGTase were pooled, dialyzed against 10 mM acetate buffer (pH 5.5), and concentrated in an ultrafiltration cell (Amicon, Merck KGaA, Darmstadt, Germany) fitted with a 10 kDa membrane. The concentrate was then loaded onto a column of Sepharose activated with α-cyclodextrin and equilibrated with the acetate buffer. After washing, the proteins were eluted with the above buffer containing α-cyclodextrin (10 mg/mL). CGTase-enriched fractions were merged, concentrated as before and stored at −20 °C until its use. Purity was assessed by Coomasie-Blue stained SDS-PAGE electrophoresis.

3.9. Critical Micellar Concentration

The measurement of the surface tension in H_2O was carried out at 20 °C using different aqueous solutions of alkyl glycosides at various concentrations, according to the Wilhelmy plate method [56] with a tensiometer (Processor tensiometer K-12, Krüss, Hamburg, Germany), which measures the real tension values at the equilibrium. Critical micelle concentrations (CMCs) were calculated graphically from the sharp change in the slope of the surface tension values vs. the logarithm of surfactant concentration, expressed in mM.

The average area (A) occupied per molecule of surfactant adsorbed in the saturated water-air interface was estimated from the equation

$$A = \frac{10^{16}}{N_A} \times \Gamma_m$$

in which N_A is the Avogadro's number, Γ_m is the maximum concentration of surfactant molecules adsorbed in the saturated interface (moles/cm^2), and the resulting A is expressed in squared Angstroms. The value of Γ_m can be calculated by applying the Gibbs equation:

$$\Gamma_m = -\frac{\left(\frac{d\gamma}{d\log c}\right)}{2.3030 \cdot n \cdot R \cdot T}$$

where ($d\gamma$/d log c) is the maximum slope of the linear plot representing surface tension vs. logarithm of surfactant concentration that appears immediately below the CMC, R = 8.31 J·mol^{-1}·K^{-1}, and T is the temperature in K. The value of n (the number of species into which the surfactant dissociates) is taken as one for nonionic surfactants.

4. Conclusions

The present work describes, for the first time, the enzymatic synthesis of α-glucosides of a tertiary alcohol (*tert*-butyl alcohol). The process is catalyzed by the enzyme CGTase, in particular from the strains *Thermoanaerobacter* sp. and *T. thermosulfurigenes*. The biotransformation is carried out under mild conditions, using starch as glucose donor, and gives rise to a significant yield (44%) of α-glucoside (major product) and α-maltoside (minor product). The synthesized *tert*-butyl-O-α-D-glucopyranoside exhibited the typical surfactant behavior and its properties were comparable to those of the related α-octyl derivative.

Supplementary Materials: The following are available online at http://www.mdpi.com/2073-4344/9/7/575/s1, Figure S1: ESI-TOF spectrum (positive mode) of *tert*-butyl-O-α-D-glucopyranoside, Figure S2: ESI-TOF spectrum (positive mode) of *tert*-butyl-O-α-D-maltoside.

Author Contributions: F.J.P., M.A. and A.O.B. conceived and designed the experiments; H.G.-A., J.L.G.-A. and C.U. carried out most of the experiments; M.B. performed the product characterization by NMR; F.C. and J. L.P. contributed with surfactant analysis; and F.J.P. and H.G.-A. wrote the paper, which was improved by the rest of authors.

Funding: This work was supported by a grant from the Spanish Ministry of Economy and Competitiveness (Grants BIO2016-76601-C3-1-R). We thank the support of the EU COST-Action CM1303 on Systems Biocatalysis. We thank Spanish Ministry of Education for a FPU grant to J.L.G.-A.

Acknowledgments: We thank L. Dijkhuizen (Groningen University, The Netherlands) for providing us CGTases from *Thermoanaerobacterium thermosulfurigenes* EM1 and *Bacillus circulans*. We are grateful to Yoshihiko Hirose (Enzyme Techno) for the supply of CGTase from *Geobacillus* sp. We thank Ramiro Martinez (Novozymes A/S) for supplying Toruzyme and for significant suggestions. We acknowledge support of the publication fee by the CSIC Open Access Publication Support Initiative through its Unit of Information Resources for Research (URICI).

Conflicts of Interest: The authors declare no conflict of interest.

References

1. Uitdehaag, J.C.M.; Van Der Veen, B.A.; Dijkhuizen, L.; Dijkstra, B.W. Catalytic mechanism and product specificity of cyclodextrin glycosyltransferase, a prototypical transglycosylase from the α-amylase family. *Enzyme Microb. Technol.* **2002**, *30*, 295–304. [CrossRef]
2. Leemhuis, H.; Kelly, R.M.; Dijkhuizen, L. Engineering of cyclodextrin glucanotransferases and the impact for biotechnological applications. *Appl. Microbiol. Biotechnol.* **2010**, *85*, 823–835. [CrossRef] [PubMed]
3. Fenyvesi, É.; Vikmon, M.; Szente, L. Cyclodextrins in Food Technology and Human Nutrition: Benefits and Limitations. *Crit. Rev. Food Sci. Nutr.* **2016**, *56*, 1981–2004. [CrossRef] [PubMed]
4. Alcalde, M.; Plou, F.J.; Andersen, C.; Martin, M.T.; Pedersen, S.; Ballesteros, A. Chemical modification of lysine side chains of cyclodextrin glycosyltransferase from *Thermoanaerobacter* causes a shift from cyclodextrin glycosyltransferase to alpha-amylase specificity. *FEBS Lett.* **1999**, *445*, 333–337. [CrossRef]
5. Wind, R.D.; Buitelaar, R.M.; Dijkhuizen, L. Engineering of factors determining alpha-amylase and cyclodextrin glycosyltransferase specificity in the cyclodextrin glycosyltransferase from *Thermoanaerobacterium thermosulfurigenes* EM1. *Eur. J. Biochem.* **1998**, *253*, 598–605. [CrossRef]
6. Tonkova, A. Bacterial cyclodextrin glucanotransferase. *Enzyme Microb. Technol.* **1998**, *22*, 678–686. [CrossRef]
7. Alcalde, M.; Plou, F.J.; Martin, M.T.; Valdes, I.; Mendez, E.; Ballesteros, A. Succinylation of cyclodextrin glycosyltransferase from *Thermoanaerobacter* sp. 501 enhances its transferase activity using starch as donor. *J. Biotechnol.* **2001**, *86*, 71–80. [CrossRef]
8. Martin, M.T.; Cruces, M.A.; Alcalde, M.; Plou, F.J.; Bernabe, M.; Ballesteros, A. Synthesis of maltooligosyl fructofuranosides catalyzed by immobilized cyclodextrin glucosyltransferase using starch as donor. *Tetrahedron* **2004**, *60*, 529–534. [CrossRef]
9. Torres, P.; Poveda, A.; Jimenez-Barbero, J.; Parra, J.L.; Comelles, F.; Ballesteros, A.O.; Plou, F.J. Enzymatic synthesis of α-glucosides of resveratrol with surfactant activity. *Adv. Synth. Catal.* **2011**, *353*, 1077–1086. [CrossRef]
10. Mathew, S.; Adlercreutz, P. Regioselective glycosylation of hydroquinone to α-arbutin by cyclodextrin glucanotransferase from *Thermoanaerobacter* sp. *Biochem. Eng. J.* **2013**, *79*, 187–193. [CrossRef]

11. Monthieu, C.; Guibert, A.; Taravel, F.R.; Nardin, R.; Combes, D. Purification and characterisation of polyglucosyl-fructosides produced by means of cyclodextrin glucosyl transferase. *Biocatal. Biotransf.* **2003**, *21*, 7–15. [CrossRef]

12. Martin, M.T.; Alcalde, M.; Plou, F.J.; Dijkhuizen, L.; Ballesteros, A. Synthesis of malto-oligosaccharides via the acceptor reaction catalyzed by cyclodextrin glycosyltransferases. *Biocatal. Biotransf.* **2001**, *19*, 21–35. [CrossRef]

13. Han, R.; Ge, B.; Jiang, M.; Xu, G.; Dong, J.; Ni, Y. High production of genistein diglucoside derivative using cyclodextrin glycosyltransferase from *Paenibacillus macerans*. *J. Ind. Microbiol. Biotechnol.* **2017**, *44*, 1343–1354. [CrossRef] [PubMed]

14. Gonzalez-Alfonso, J.L.; Leemans, L.; Poveda, A.; Jiménez-Barbero, J.; Ballesteros, A.O.; Plou, F.J. Efficient α-glucosylation of epigallocatechin gallate catalyzed by cyclodextrin glucanotransferase from *Thermoanaerobacter* sp. *J. Agric. Food Chem.* **2018**, *66*, 7402–7408. [CrossRef] [PubMed]

15. Jun, H.K.; Bae, K.M.; Kim, S.K. Production of 2-O-alpha-D-glucopyranosyl L-ascorbic acid using cyclodextrin glucanotransferase from *Paenibacillus* sp. *Biotechnol. Lett.* **2001**, *23*, 1793–1797. [CrossRef]

16. Miranda-Molina, A.; Marquina-Bahena, S.; López-Munguía, A.; Álvarez, L.; Castillo, E. Regioselective glucosylation of inositols catalyzed by *Thermoanaerobacter* sp. CGTase. *Carbohydr. Res.* **2012**, *360*, 93–101. [CrossRef] [PubMed]

17. Nakano, H.; Kitahata, S. Application of cyclodextrin glucanotransferase to the synthesis of useful oligosaccharides and glycosides. In *Handbook of Industrial Biocatalysts*; Hou, C.T., Ed.; Taylor and Francis: Boca Raton, FL, USA, 2005; pp. 419–435.

18. Do, H.; Sato, T.; Kirimura, K.; Kino, K.; Usami, S. Enzymatic synthesis of L-menthyl alpha-maltoside and L-menthyl alpha-maltooligosides from L-menthyl alpha-glucoside by cyclodextrin glucanotransferase. *J. Biosci. Bioeng.* **2002**, *94*, 119–123. [CrossRef]

19. Kelly, R.M.; Dijkhuizen, L.; Leemhuis, H. The evolution of cyclodextrin glucanotransferase product specificity. *Appl. Microbiol. Biotechnol.* **2009**, *84*, 119–133. [CrossRef]

20. Kurimoto, M.; Tabuchi, A.; Mandai, T.; Shibuya, T.; Chaen, H.; Fukuda, S.; Sugimoto, T.; Tsujisaka, Y. Synthesis of glycosyl-trehaloses by cyclomaltodextrin glucanotransferase through the transglycosylation reaction. *Biosci. Biotechnol. Biochem.* **1997**, *61*, 1146–1149. [CrossRef]

21. Shimoda, K.; Hara, T.; Hamada, H.; Hamada, H. Synthesis of curcumin β-maltooligosaccharides through biocatalytic glycosylation with *Strophanthus gratus* cell culture and cyclodextrin glucanotransferase. *Tetrahedron Lett.* **2007**, *48*, 4029–4032. [CrossRef]

22. González-Alfonso, J.; Rodrigo-Frutos, D.; Belmonte-Reche, E.; Peñalver, P.; Poveda, A.; Jiménez-Barbero, J.; Ballesteros, A.O.; Hirose, Y.; Polaina, J.; Morales, J.; et al. Enzymatic synthesis of a novel pterostilbene α-glucoside by the combination of cyclodextrin glucanotransferase and amyloglucosidase. *Molecules* **2018**, *23*, 1271. [CrossRef] [PubMed]

23. González-Alfonso, J.L.; Míguez, N.; Padilla, J.D.; Leemans, L.; Poveda, A.; Jiménez-Barbero, J.; Ballesteros, A.O.; Sandoval, G.; Plou, F.J. Optimization of regioselective α-glucosylation of hesperetin catalyzed by cyclodextrin glucanotransferase. *Molecules* **2018**, *23*, 2885. [CrossRef] [PubMed]

24. Han, R.; Liu, L.; Shin, H.; Chen, R.R.; Du, G.; Chen, J. Site-saturation engineering of lysine 47 in cyclodextrin glycosyltransferase from *Paenibacillus macerans* to enhance substrate specificity towards maltodextrin for enzymatic synthesis of 2-O-D-glucopyranosyl-L-ascorbic acid (AA-2G). *Appl. Microbiol. Biotechnol.* **2013**, *97*, 5851–5860. [CrossRef] [PubMed]

25. Wind, R.D.; Uitdehaag, J.C.M.; Buitelaar, R.M.; Dijkstra, B.W.; Dijkhuizen, L. Engineering of cyclodextrin product specificity and pH optima of the thermostable cyclodextrin glycosyltransferase from *Thermoanaerobacterium thermosulfurigenes* EM1. *J. Biol. Chem.* **1998**, *273*, 5771–5779. [CrossRef] [PubMed]

26. Koo, Y.S.; Lee, H.W.; Jeon, H.Y.; Choi, H.J.; Choung, W.J.; Shim, J.H. Development and characterization of cyclodextrin glucanotransferase as a maltoheptaose-producing enzyme using site-directed mutagenesis. *Protein Eng. Des. Sel.* **2015**, *28*, 531–537. [CrossRef] [PubMed]

27. Martin, M.T.; Plou, F.J.; Alcalde, M.; Ballesteros, A. Immobilization on Eupergit C of cyclodextrin glucosyltransferase (CGTase) and properties of the immobilized biocatalyst. *J. Mol. Catal. B Enzym.* **2003**, *21*, 299–308. [CrossRef]

28. Iyer, J.L.; Shetty, P.; Pai, J.S. Immobilisation of cyclodextrin glucanotransferase from *Bacillus circulans* ATCC 21783 on purified seasand. *J. Ind. Microbiol. Biotechnol.* **2003**, *30*, 47–51. [CrossRef]

29. Delbourg, M.F.; Drouet, P.; Demoraes, F.; Thomas, D.; Barbotin, J.N. Effect of PEG and other additives on cyclodextrin production by *Bacillus macerans* cyclomaltodextrin-glycosyl-transferase. *Biotechnol. Lett.* **1993**, *15*, 157–162. [CrossRef]

30. Morita, T.; Yoshida, N.; Karube, I. A novel synthesis method for cyclodextrins from maltose in water-organic solvent systems. *Appl. Biochem. Biotechnol.* **1996**, *56*, 311–324. [CrossRef]

31. Blackwood, A.D.; Bucke, C. Addition of polar organic solvents can improve the product selectivity of cyclodextrin glycosyltransferase-Solvent effects on CGTase. *Enzyme Microb. Technol.* **2000**, *27*, 704–708. [CrossRef]

32. Qi, Q.S.; Zimmermann, W. Cyclodextrin glucanotransferase: From gene to applications. *Appl. Microbiol. Biotechnol.* **2005**, *66*, 475–485. [CrossRef] [PubMed]

33. van Rantwijk, F.; Oosterom, M.W.V.; Sheldon, R.A. Glycosidase-catalysed synthesis of alkyl glycosides. *J. Mol. Catal. B Enzym.* **1999**, *6*, 511–532. [CrossRef]

34. Svasti, J.; Phongsak, T.; Sarnthima, R. Transglucosylation of tertiary alcohols using cassava beta-glucosidase. *Biochem. Biophys. Res. Commun.* **2003**, *305*, 470–475. [CrossRef]

35. Huneke, F.U.; Bailey, D.; Nucci, R.; Cowan, D. *Sulfolobus solfataricus* β-glycosidase-catalysed synthesis of sugar-alcohol conjugates in the presence of organic solvents. *Biocatal. Biotransf.* **2000**, *18*, 291–299. [CrossRef]

36. Mazzaferro, L.S.; Weiz, G.; Braun, L.; Kotik, M.; Pelantová, H.; Křen, V.; Breccia, J.D. Enzyme-mediated transglycosylation of rutinose (6-O-α-L-rhamnosyl-D-glucose) to phenolic compounds by a diglycosidase from *Acremonium* sp. DSM 24697. *Biotechnol. Appl. Biochem.* **2019**, *66*, 53–59. [CrossRef] [PubMed]

37. Yang, J.; Pérez, B.; Anankanbil, S.; Li, J.; Gao, R.; Guo, Z. Enhanced synthesis of alkyl galactopyranoside by *Thermotoga naphthophila* β-galactosidase catalyzed transglycosylation: Kinetic insight of a functionalized ionic liquid-mediated system. *ACS Sustain. Chem. Eng.* **2017**, *5*, 2006–2014. [CrossRef]

38. García, C.; Hoyos, P.; Hernáiz, M.J. Enzymatic synthesis of carbohydrates and glycoconjugates using lipases and glycosidases in green solvents. *Biocatal. Biotransf.* **2018**, *36*, 131–140. [CrossRef]

39. von Rybinski, W.; Hill, K. Alkyl polyglycosides-Properties and applications of a new class of surfactants. *Angew. Chem. Int. Ed.* **1998**, *37*, 1328–1345. [CrossRef]

40. Sarney, D.B.; Vulfson, E.N. Application of enzymes to the synthesis of surfactants. *Trends Biotechnol.* **1995**, *13*, 164–172. [CrossRef]

41. Gonzalez-Alfonso, J.L.; Peñalver, P.; Ballesteros, A.O.; Morales, J.C.; Plou, F.J. Effect of α-glucosylation on the stability, antioxidant properties, toxicity, and neuroprotective activity of (–)-epigallocatechin gallate. *Front. Nutr.* **2019**, *6*. [CrossRef]

42. Plou, F.J.; Martin, M.T.; Gomez de Segura, A.; Alcalde, M.; Ballesteros, A. Glucosyltransferases acting on starch or sucrose for the synthesis of oligosaccharides. *Can. J. Chem.* **2002**, *80*, 743–752. [CrossRef]

43. Svensson, D.; Ulvenlund, S.; Adlercreutz, P. Efficient synthesis of a long carbohydrate chain alkyl glycoside catalyzed by cyclodextrin glycosyltransferase (CGTase). *Biotech. Bioeng.* **2009**, *104*, 854–861. [CrossRef] [PubMed]

44. Nakano, H.; Kitahata, S.; Tominaga, Y.; Kiku, Y.; Ando, K.; Kawashima, Y.; Takenishi, S. Syntheses of glucosides with trimethylolpropane and 2 related polyol moieties by cyclodextrin glucanotransferase and their esterification by lipase. *J. Ferment. Bioeng.* **1992**, *73*, 237–238. [CrossRef]

45. Larsson, J.; Svensson, D.; Adlercreutz, P. α-Amylase-catalysed synthesis of alkyl glycosides. *J. Mol. Catal. B Enzym.* **2005**, *37*, 84–87. [CrossRef]

46. Otto, R.T.; Bornscheuer, U.T.; Syldatk, C.; Schmid, R.D. Synthesis of aromatic n-alkyl-glucoside esters in a coupled β-glucosidase and lipase reaction. *Biotechnol. Lett.* **1998**, *20*, 437–440. [CrossRef]

47. Fischer, L.; Bromann, R.; Kengen, S.W.M.; de Vos, W.M.; Wagner, F. Catalytical potency of β-glucosidase from the extremophile *Pyrococcus furiosus* in glucoconjugate synthesis. *Bio/Technology* **1996**, *14*, 88–91. [CrossRef]

48. Jiang, Z.; Zhu, Y.; Li, L.; Yu, X.; Kusakabe, I.; Kitaoka, M.; Hayashi, K. Transglycosylation reaction of xylanase B from the hyperthermophilic *Thermotoga maritima* with the ability of synthesis of tertiary alkyl β-D-xylobiosides and xylosides. *J. Biotechnol.* **2004**, *114*, 125–134. [CrossRef]

49. Kongsaeree, P.T.; Ratananikom, K.; Choengpanya, K.; Tongtubtim, N.; Sujiwattanarat, P.; Porncharoennop, C.; Onpium, A.; Svasti, J. Substrate specificity in hydrolysis and transglucosylation by family 1 β-glucosidases from cassava and Thai rosewood. *J. Mol. Catal. B Enzym.* **2010**, *67*, 257–265. [CrossRef]

50. Simerská, P.; Kuzma, M.; Monti, D.; Riva, S.; Macková, M.; Křen, V. Unique transglycosylation potential of extracellular α-D-galactosidase from *Talaromyces flavus*. *J. Mol. Catal. B Enzym.* **2006**, *39*, 128–134. [CrossRef]

51. Tanaka, T.; Kikuta, N.; Kimura, Y.; Shoda, S.I. Metal-catalyzed stereoselective and protecting-group-free synthesis of 1,2-cis-glycosides using 4,6-dimethoxy-1,3,5-triazin-2-yl glycosides as glycosyl donors. *Chem. Lett.* **2015**, *44*, 846–848. [CrossRef]
52. Zamost, B.L.; Nielsen, H.K.; Starnes, R.L. Thermostable enzymes for industrial applications. *J. Ind. Microbiol.* **1991**, *8*, 71–81. [CrossRef]
53. De Roode, B.M.; Franssen, M.C.R.; Van Der Padt, A.; Boom, R.M. Perspectives for the industrial enzymatic production of glycosides. *Biotechnol. Prog.* **2003**, *19*, 1391–1402. [CrossRef] [PubMed]
54. Wasylewski, Z.; Kozik, A. Protein-nonionic detergent interaction-interaction of bovine serum-albumin with alkyl glucosides studied by equilibrium dialysis and infrared spectroscopy. *Eur. J. Biochem.* **1979**, *95*, 121–126. [CrossRef] [PubMed]
55. Nakamura, A.; Haga, K.; Yamane, K. The transglycosylation reaction of cyclodextrin glucanotransferase is operated by a ping-pong mechanism. *FEBS Lett.* **1994**, *337*, 66–70. [CrossRef]
56. International Standard ISO 304-1985. Surface active agents. Determination of surface tension by drawing up liquid films. Available online: https://www.iso.org/standard/4238.html (accessed on 29 June 2019).

Article

Production of New Isoflavone Glucosides from Glycosylation of 8-Hydroxydaidzein by Glycosyltransferase from *Bacillus subtilis* ATCC 6633

Chien-Min Chiang [1], Tzi-Yuan Wang [2], Szu-Yi Yang [3], Jiumn-Yih Wu [4,*] and Te-Sheng Chang [3,*]

[1]   Department of Biotechnology, Chia Nan University of Pharmacy and Science, No. 60, Sec. 1, Erh-Jen Rd., Jen-Te District, Tainan 71710, Taiwan; cmchiang@mail.cnu.edu.tw
[2]   Biodiversity Research Center, Academia Sinica, Taipei 115, Taiwan; tziyuan@gmail.com
[3]   Department of Biological Sciences and Technology, National University of Tainan, Tainan 70005, Taiwan; szuyi08231995@gmail.com
[4]   Department of Food Science, National Quemoy University, Kinmen County 892, Taiwan
*   Correspondence: wujy@nqu.edu.tw (J.-Y.W.); mozyme2001@gmail.com (T.-S.C.); Tel.: +886-8231-3310 (J.-Y.W.); Fax: +886-8231-3797 (J.-Y.W.); Tel./Fax: +886-6260-2137 (T.-S.C.)

Received: 24 August 2018; Accepted: 7 September 2018; Published: 10 September 2018

Abstract: 8-Hydroxydaidzein (8-OHDe) has been proven to possess some important bioactivities; however, the low aqueous solubility and stability of 8-OHDe limit its pharmaceutical and cosmeceutical applications. The present study focuses on glycosylation of 8-OHDe to improve its drawbacks in solubility and stability. According to the results of phylogenetic analysis with several identified flavonoid-catalyzing glycosyltransferases (GTs), three glycosyltransferase genes (*BsGT110*, *BsGT292* and *BsGT296*) from the genome of the *Bacillus subtilis* ATCC 6633 strain were cloned and expressed in *Escherichia coli*. The three BsGTs were then purified and the glycosylation activity determined toward 8-OHDe. The results showed that only *BsGT110* possesses glycosylation activity. The glycosylated metabolites were then isolated with preparative high-performance liquid chromatography and identified as two new isoflavone glucosides, 8-OHDe-7-*O*-β-glucoside and 8-OHDe-8-*O*-β-glucoside, whose identity was confirmed by mass spectrometry and nuclear magnetic resonance spectroscopy. The aqueous solubility of 8-OHDe-7-*O*-β-glucoside and 8-OHDe-8-*O*-β-glucoside is 9.0- and 4.9-fold, respectively, higher than that of 8-OHDe. Moreover, more than 90% of the initial concentration of the two 8-OHDe glucoside derivatives remained after 96 h of incubation in 50 mM of Tris buffer at pH 8.0. In contrast, the concentration of 8-OHDe decreased to 0.8% of the initial concentration after 96 h of incubation. The two new isoflavone glucosides might have potential in pharmaceutical and cosmeceutical applications.

Keywords: *Bacillus*; glycosyltransferase; 8-hydroxydaidzein

---

1. Introduction

Daidzein and genistein, the major isoflavones found in soybean, have been under intensive investigation in the past few decades due to their potential roles in preventing certain hormone-dependent diseases [1]. In recent years, biotransformation of isoflavones using either wild-type or genetically-engineered microorganisms has also been of interest because the bioactivity of isoflavones dramatically alters after biotransformation. Among the various biotransformations of soy isoflavones, *ortho*-hydroxylation of soy isoflavones has become a subject of great interest because of the *ortho*-hydroxydaidzein and *ortho*-hydroxygenistein derivatives produced, which usually possess higher bioactivities compared with those of the precursors, daidzein and genistein [2].

8-Hydroxydaidzein (8-OHDe) is one of the *ortho*-hydroxydaidzein derivatives. The compound can be produced from biotransformation of daidzein by either wild *Aspergillus oryzae* [3–5], genetically-engineered *Pichia pastoris* [6] or *Escherichia coli* [7,8]. Multiple bioactivities related to 8-OHDe have been reported, including anti-cancer [9], suppression of multidrug resistance [10], anti-tyrosinase [11,12], skin whitening [13,14], anti-aldose reductase [15] and the newly-found anti-inflammatory activity [16,17].

Although 8-OHDe has been identified with many bioactivities, some drawbacks (including low aqueous solubility and stability) limit isoflavone's application in pharmaceuticals and cosmeceuticals [18]. To improve the drawbacks of 8-OHDe, biotransformation through whole cells or enzyme biocatalysts into 8-OHDe derivatives is a promising strategy. Among various biotransformations, glycosylation, the attachment of a bulky sugar group to the precursor molecules, could improve the chemical stability and aqueous solubility of natural compounds. The aqueous solubility of the corresponding 7-*O*-glucoside of soy isoflavones is improved about 30-fold [19]. Increasing the aqueous solubility and stability could expand applications of the compounds in advance. Therefore, in the present study, we are interested in investigating the glycosyl-biotransformation of 8-OHDe.

Glycosylation is a common modification reaction in the biosynthesis of natural compounds. Generally, glycosylation is catalyzed by glycosyltransferases (GTs, EC 2.4.x.y), which transfer sugar moieties from the activated donor molecules to specific acceptor molecules [20–22]. Our previous studies found that *Bacillus subtilis* ATCC 6633 could biotransform antcin K, which is a major ergostane triterpenoid from the fruiting bodies of *Antrodia cinnamomea*, to its glucoside derivatives [23]. In the present study, the *B. subtilis* strain was found to biotransform 8-OHDe. To identify the biotransformation in advance, three GT genes were cloned from the *B. subtilis* strain and overexpressed in *Escherichia coli*. Then, the biotransformation activity of the three purified GT enzymes toward 8-OHDe was determined. The biotransformed metabolites by one positive-active enzyme were isolated and identified. Finally, the aqueous solubility and stability of the 8-OHDe glucoside derivatives were determined.

2. Results and Discussion

2.1. Confirming Biotransformation of 8-OHDe by Bacillus subtilis ATCC 6633

Our previous studies showed that *B. subtilis* ATCC 6633 could biotransform the ergostane triterpenoid antcin K to its glucoside derivatives [23]. To confirm whether *B. subtilis* ATCC 6633 could also biotransform 8-OHDe, the bacterium was cultivated in broth with 8-OHDe, and the fermentation broth was analyzed using ultra-performance liquid chromatography (UPLC).

Figure 1 shows the UPLC analysis of the initial (dashed line) and 24-h (solid line) fermentation broths of the strain *B. subtilis* ATCC 6633 fed with 8-OHDe. In the figure, 8-OHDe appears at the retention time (RT) of 5.1 min. After 24 h of fermentation, the peak of the precursor decreases, while several new peaks with RTs of between 3 and 4 min appear. To confirm whether the new peaks are from biotransformation of 8-OHDe, biotransformation was conducted in the absence of 8-OHDe. The results showed that the new peaks did not appear at 24 h of the fermentation broth of the strain in the absence of 8-OHDe (Figure S1). Thus, it was concluded that 8-OHDe was biotransformed by the strain *B. subtilis* ATCC 6633.

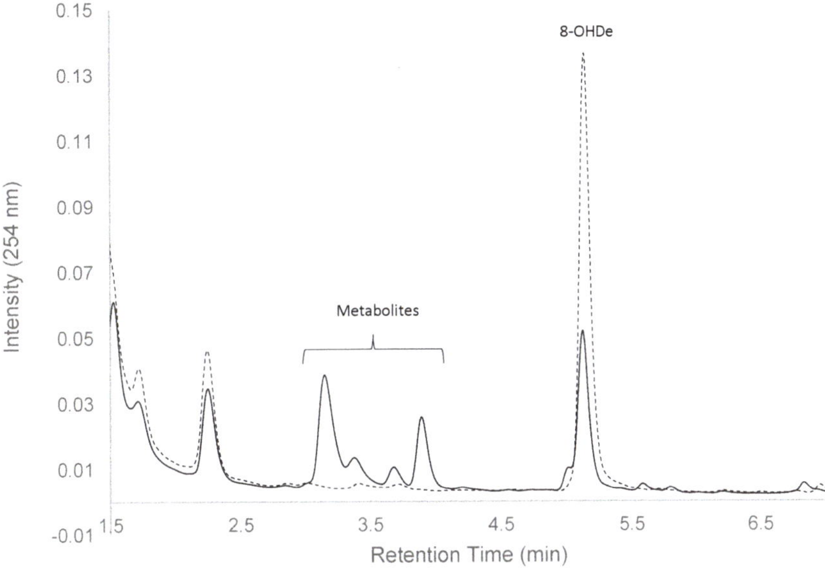

Figure 1. Biotransformation of 8-hydroxydaidzein (8-OHDe) by *B. subtilis* ATCC 6633. The strain was cultivated in modified glucose nutrient (MGN) media containing 0.02 mg/mL of 8-OHDe. The initial (dashed line) and 24-h (solid line) cultivations of the fermentation broth were analyzed with UPLC. The UPLC operation conditions are described in the Materials and Methods.

2.2. Phylogenetic Analysis of GTs from B. subtilis ATCC 6633

Although *B. subtilis* ATCC 6633 has been proven to biotransform the ergostane triterpenoid antcin K [23] and 8-OHDe (Figure 1), biotransformation by using whole cells as biocatalysts usually has lower efficiency than that by using purified enzyme, which can be produced through genetic engineering. Thus, cloning of the putative genes encoding the catalytic enzymes in the biotransformation is a worthy strategy. According to the genome data of *B. subtilis* ATCC 6633 (GenBank BioProject Accession No. PRJNA43011), there are 28 GTs annotated in the genome. To clone the GT genes from *B. subtilis* ATCC 6633 responsible for the biotransformation of 8-OHDe, phylogenetic analysis of the annotated GTs from *B. subtilis* ATCC 6633 was compared with known bacterial GTs, which have been proven to possess glycosylation activity toward flavonoids. The characterized bacterial glycosyltransferases included BlYjiC (AAU40842) from *B. licheniformis* ATCC 14580 [24], BsYjiC (NP_389104) from *B. subtilis* 168 [25,26], BcGT-1 (AAS41089) and BcGT-3 (AAS41737) from *B. cereus* ATCC 10987 [27,28], OleD (ABA42119) from *Streptomyces antibioticus* [29] and XcGT-2 (AAM41712) from *Xanthomonas campestris* pv. *campestris* ATCC 33913 [30]. The results are shown in Figure 2. According to the results of the phylogenetic analysis, all tested six identified flavonoid-catalyzing GTs (gray background in Figure 2) were clustered into one group (Figure 2). Among the 28 BsGTs, only *BsGT110* was included in the group. At the same time, *BsGT292* and *BsGT296* were close to the group. Thus, the three GTs, *BsGT110* (GenBank Protein Accession No. WP_003220110), *BsGT292* (GenBank Protein Accession No. WP_032727292) and *BsGT296* (GenBank Protein Accession No. WP_003219296; bold text in Figure 2), were selected for further functional assay.

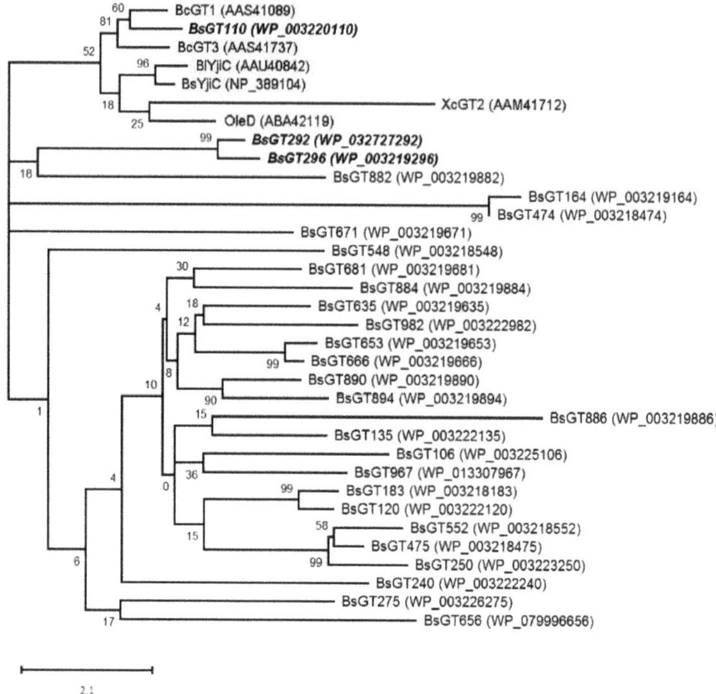

Figure 2. Unrooted phylogenetic analysis of glycosyltransferase genes by using the maximum likelihood method. The tree with the highest log likelihood (−19,549.15) is shown based on the general reversible mitochondrial model [31]. The percentage of trees in which the associated taxa clustered together is shown next to the branches. Initial trees for the heuristic search were obtained automatically by applying the neighbor-join and BioNJ algorithms to a matrix of pairwise distances estimated using a JTT model and then selecting the topology with the superior log likelihood value. The rate variation model allowed for some sites to be evolutionarily invariable ([+I], 0.00% sites). The tree is drawn to scale, with branch lengths measured in the number of substitutions per site. The analysis involved 34 protein sequences. All positions with less than 95% site coverage were eliminated. That is, fewer than 5% alignment gaps, missing data and ambiguous bases were allowed at any position. There were a total of 210 positions in the final dataset. Evolutionary analyses were conducted in Molecular Evolutionary Genetics Analysis (MEGA X) across computing platforms (Version 10.0.4, Center for Evolutionary Functional Genomics, The Biodesign Institute, Arizona State University, Tempe, AZ, USA, 1993–2018) [32].

2.3. Cloning, Overexpression, Purification and Activity Assay of BsGTs from B. subtilis ATCC 6633 in E. coli

To identify which enzyme catalyzed the biotransformation, the three BsGT genes were cloned into a pETDuet-1 expression vector (Figure 3a). The recombinant BsGT gene in recombinant *E. coli* was overexpressed by induction with 0.2 mM of isopropyl β-D-1-thiogalactopyranoside (IPTG), and the protein produced was purified with Ni^{2+} chelate affinity chromatography. Figure 3b shows the sodium dodecyl sulfate polyacrylamide gel electrophoresis (SDS-PAGE) analysis of the overexpressed and purified *BsGT110*, while the SDS-PAGE analysis of purified *BsGT292* and *BsGT296* is in Figure S2.

The purified enzymes were incubated with uridine diphosphate (UDP)-glucose and the precursor 8-OHDe to confirm their biotransformation activity toward 8-OHDe. The results showed that only *BsGT110* has biotransformation activity toward 8-OHDe (Figure 4), while neither purified *BsGT292*, nor *BsGT296* showed the activity (Figure S3). The result was consistent with the phylogenetic analysis results (Figure 2), where only *BsGT110* was clustered with the group of flavonoid-catalyzing GTs.

In addition, the RTs of the two biotransformation metabolites (Figure 4), 3.7 min and 3.9 min for Compound (**1**) and Compound (**2**), respectively, were similar to some of the new peaks in the 8-OHDe biotransformation by the strain *B. subtilis* ATCC 6633 (Figure 1). The results imply that *BsGT110* is one of the corresponding enzymes catalyzing 8-OHDe during the biotransformation of 8-OHDe by the strain *B. subtilis* ATCC 6633.

By comparing the results of the biotransformations using whole cells (Figure 1) and the purified *BsGT110* (Figure 4), it is obvious that biotransformation products using the purified *BsGT110* (Compound (**1**) and Compound (**2**) in Figure 4) were more specific than those using the whole cells (metabolites between RT 3 and 4 min in Figure 1). Moreover, the catalyzing efficiency using the purified *BsGT110* (reaction time of 30 min in Figure 4) was higher than that using whole cells (reaction time of 24 h in Figure 1). Therefore, biotransformation by using the purified *BsGT110* as biocatalysts has higher efficiency and specificity than those by using whole cells.

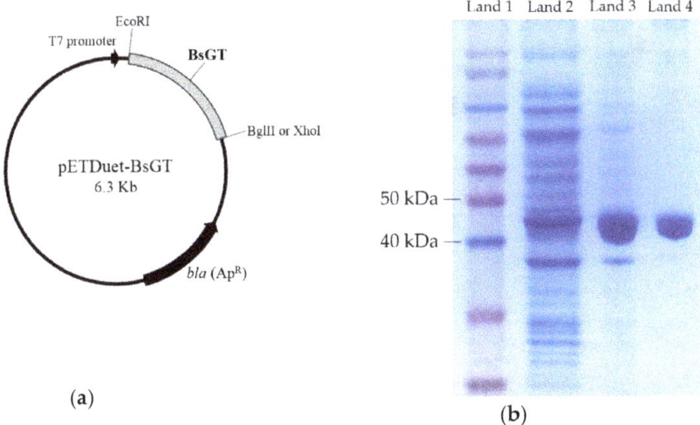

Figure 3. Expression and purification of the BsGTs from *B. subtilis* ATCC 6633 in *E. coli.* (**a**) Diagram of the recombinant expression plasmid; (**b**) SDS-PAGE analysis of expressed and purified proteins from recombinant *E. coli* harboring pETDuet-*BsGT110*. Lane 1: molecular marker; Lane 2: total protein before induction; Lane 3: total protein after 20 h induction; Lane 4: purified protein.

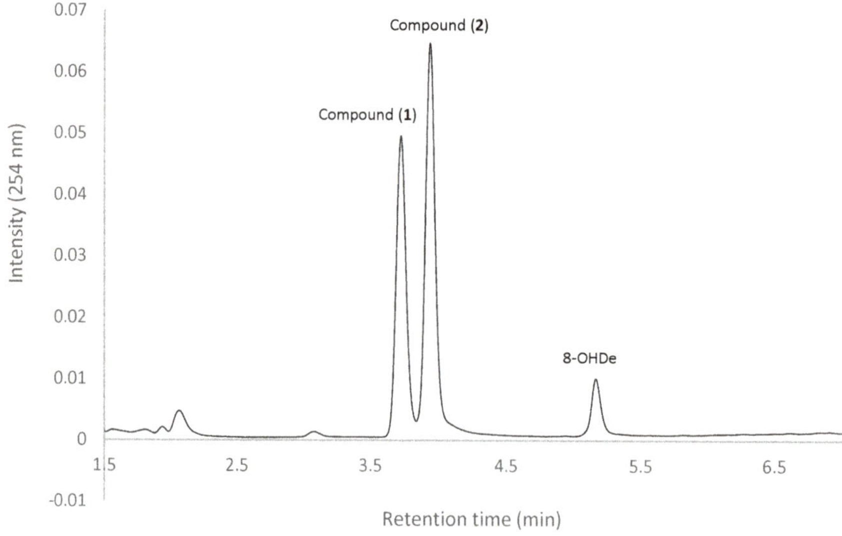

Figure 4. Biotransformation of 8-OHDe by the purified *BsGT110*. Two micrograms of the purified enzyme were incubated with 0.4 mM uridine diphosphate (UDP)-glucose and 0.02 mg/mL of 8-OHDe in the presence of 50 mM Tris at pH 8.0 and 10 mM of $MgCl_2$ at 40 °C for 30 min. After the reaction, the mixtures were analyzed with ultra-performance liquid chromatography (UPLC). The UPLC operation conditions are described in the Materials and Methods.

2.4. Optimal Catalyzing Conditions for BsGT110

To determine the optimal catalyzing condition, the activity of *BsGT110* at different pH values, temperatures and metal ions was examined. The results are shown in Figure 5. The enzyme activity was very sensitive to pH, and the enzyme activity at pH 7.0 was nearly 3–4-fold higher than that at pH 8.0 and pH 6.0. However, the enzyme activity was insensitive to temperature, and the enzyme activity was not significantly different between 30 °C and 50 °C. In addition, the enzyme favored Ca^{2+} as its cofactor. In the presence of Ca^{2+}, the enzyme activity was 1.8-fold higher than that at Mg^{2+} or none. It is known that GTs utilize divalent metal ion cofactors such as Mn^{2+} and Mg^{2+}. However, Li et al. found that the activity of the GT from *Bacillus circulans* was enhanced by Ca^{2+} due to an additional calcium-binding site in the structure [33]. Whether *BsGT110* favored Ca^{2+} as its cofactor due to an additional calcium-binding site needs to be studied in the future. From the results, the optimal catalyzing conditions for *BsGT110* are at 40 °C, pH 7.0 with 10 mM of Ca^{2+}.

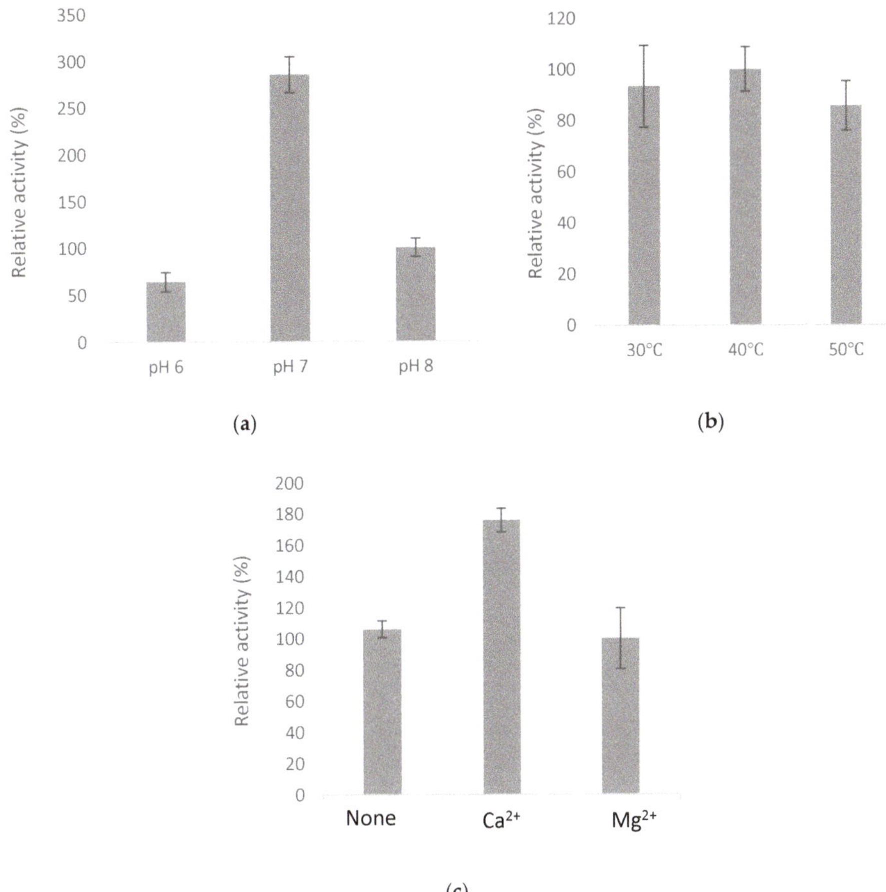

Figure 5. Effects of pH (**a**), temperature (**b**) and metal ion (**c**) on *BsGT110* activity. A standard condition was set as 2 µg of the purified *BsGT110*, 1 mg/mL of 8-OHDe, 10 mM of $MgCl_2$ and 10 mM of UDP-glucose in 50 mM of Tris at pH 8.0 and 40 °C. To determine the optimal reaction condition, the pH, temperature or metal ion in the standard condition was replaced by the tested condition. Relative activity was obtained by dividing the area of the summation of the two product peaks, Compound (**1**) and Compound (**2**), of the reaction in the UPLC profile by that of the reaction at the standard condition. The mean (n = 3) is shown, and the standard deviations are represented by error bars.

2.5. Substrate Specificity of BsGT110

To study the substrate specificity of *BsGT110*, another three *ortho*-hydroxyisoflavones (3′-hydroxydaidzein, 3′-OHDe; 3′-hydroxygenistein, 3′-OHGe; 6′-hydroxydaidzein, 6-OHDe) were used in the biotransformation assays by *BsGT110*. The result is shown in Figure 6 and reveals that *BsGT110* could also catalyze all the tested *ortho*-hydroxyisoflavones. However, unlike two major metabolites, Compound (**1**) and Compound (**2**), produced in the biotransformation of 8-OHDe by *BsGT110* (Figure 4), only one major metabolite appeared in the biotransformations of either 3′-OHDe, 3′-OHGe or 6-OHDe by *BsGT110* (Figure 6). Due to very low amounts of the three *ortho*-hydroxyisoflavones, the biotransformation metabolites from these biotransformations were not identified in advance. In addition, among the *ortho*-hydroxyisoflavones in the present study, including

3'-OHDe, 3'-OHGe, 6-OHDe and 8-OHDe, only 8-OHDe has been approved to possess bioactivity in human volunteers [13,14]. Therefore, 8-OHDe was used as a substrate for the following study.

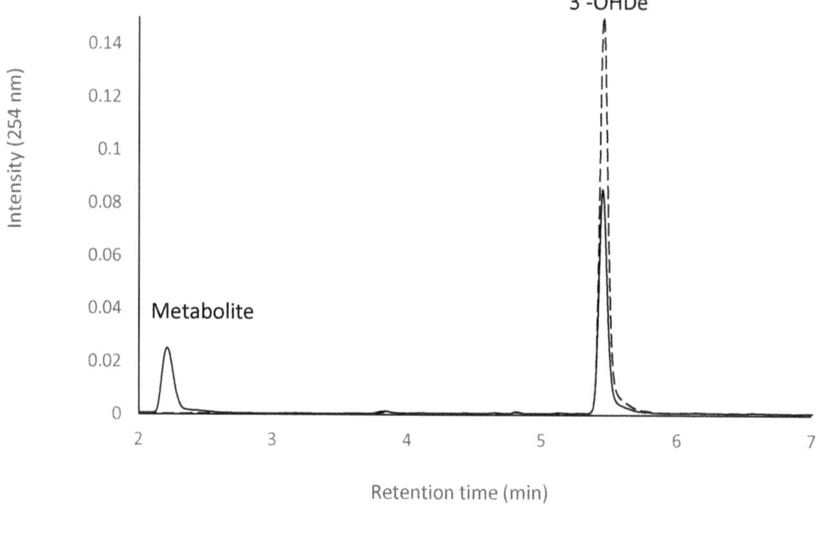

(**a**)

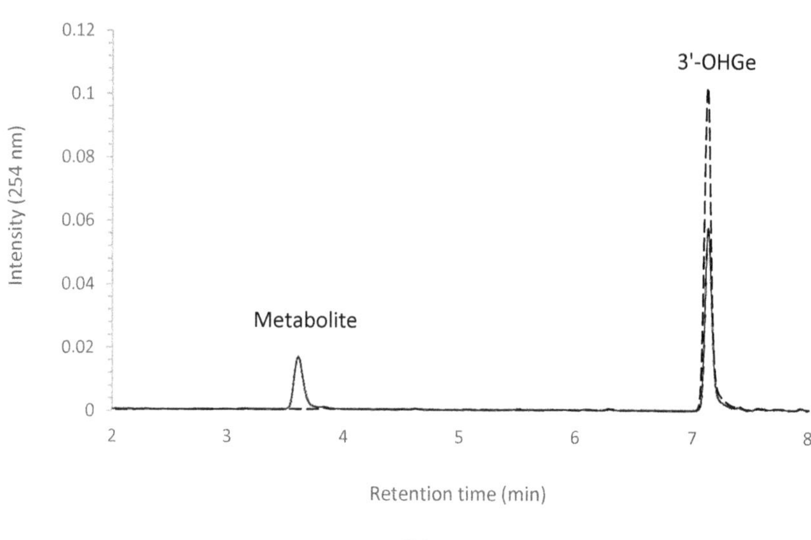

(**b**)

Figure 6. *Cont.*

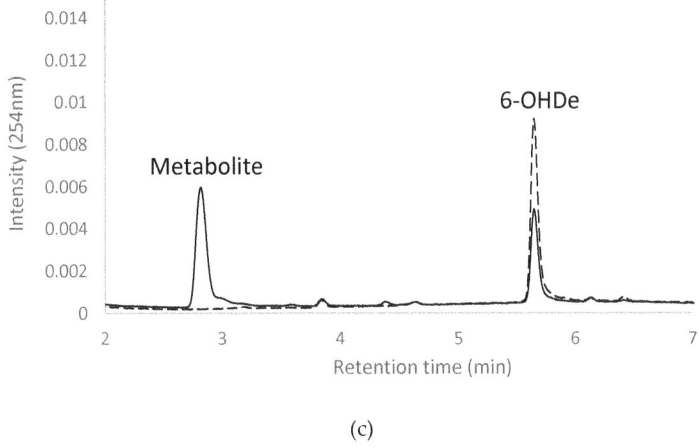

(c)

Figure 6. Biotransformation of 3′-hydroxydaidzein (3′-OHDe) (**a**), 3′-hydroxygenistein (3′-OHGe) (**b**) and 6′-hydroxydaidzein (6-OHDe) (**c**) by the purified *BsGT110*. Two micrograms of the purified enzyme were incubated with 0.4 mM uridine diphosphate (UDP)-glucose and 0.02 mg/mL of 3′-OHDe (a), 3′-OHGe (b) or 0.005 mg/mL of 6-OHDe (c) in the presence of 50 mM phosphate buffer at pH 7.0 and 10 mM of $CaCl_2$ at 40 °C for 30 min. Before (dashed line) and after (solid line) the reaction, the mixtures were analyzed with ultra-performance liquid chromatography (UPLC). The UPLC operation conditions are described in the Materials and Methods.

2.6. Stability of BsGT110

To study the stability of *BsGT110*, *BsGT110* was incubated at different temperatures for 1 h or was freeze-dried once, and then, the activity of the treated *BsGT110* was determined. The result is shown in Figure 7. The enzyme was not stable above 40 °C for 1 h, while the enzyme was stable after freeze-drying.

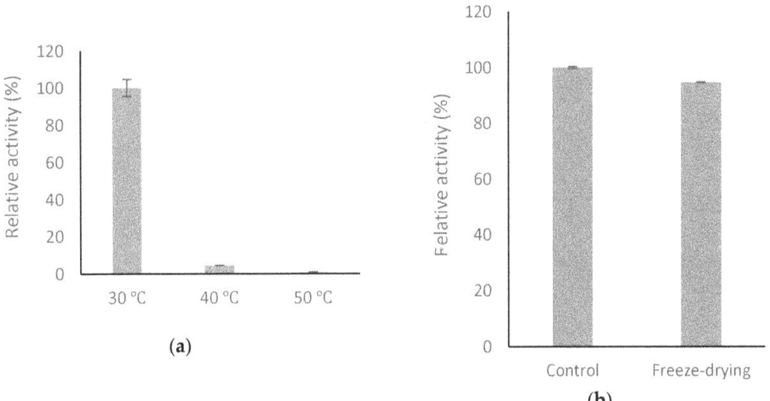

Figure 7. Stability of *BsGT110* at different temperatures (**a**) and with freeze-drying (**b**). Two micrograms of the purified *BsGT110* were pre-incubated at the tested temperature for 1 h or freeze-dried before conduction of the activity assay. A standard condition was set as 2 µg of the purified *BsGT110*, 1 mg/mL of 8-OHDe, 10 mM of $CaCl_2$ and 10 mM of UDP-glucose in 50 mM of phosphate at pH 7.0 and 40 °C. Relative activity was obtained by dividing the area of the summation of the two product peaks, Compound (**1**) and Compound (**2**), of the reaction in the UPLC profile by that of the reaction at the standard condition. The mean (*n* = 3) is shown, and the standard deviations are represented by error bars.

2.7. Isolation and Identification of Biotransformation Metabolites

To resolve the chemical structures of the metabolites, the biotransformation was scaled up at the optimal condition, and the two metabolites were purified with preparative high-performance liquid chromatography (HPLC). From a 40-mL reaction mixture containing 1 mg/mL of 8-OHDe, 10 mM of UDP-glucose, 10 mM of $CaCl_2$ and 50 mM of phosphate buffer at pH 7.0, 17.4 mg and 24.5 mg of Compound (**1**) and Compound (**2**) were isolated. Both compounds showed an [M⁻H]⁻ ion peak at m/z: 431.23 in the electrospray ionization mass (ESI-MS) spectrum corresponding to the molecular formula $C_{21}H_{20}O_{10}$. Then, ^{1}H and ^{13}C nuclear magnetic resonance (NMR), including distortionless enhancement by polarization transfer (DEPT), heteronuclear single quantum coherence (HSQC), heteronuclear multiple bond connectivity (HMBC), correlation spectroscopy (COSY) and nuclear Overhauser effect spectroscopy (NOESY) spectra, were obtained, and the ^{1}H- and ^{13}C-NMR signal assignments were conducted accordingly (shown in Figures S4–S17). In both compounds, in addition to the signals of the 8-OHDe moiety, which were confirmed by HMBC spectra (full assignments of both compounds are listed in Table S1), seven proton (from 3.18 to 4.93 ppm) and six carbon (from 60 to 105 ppm) signals corresponding to the glucose moiety structure were observed. The anomeric proton signal at δ 4.91 (J = 7.6 Hz) and 4.93 (J = 7.7 Hz) in the ^{1}H-NMR spectra of Compounds (**1**) and (**2**), respectively, indicated the β-configuration for the glucopyranosyl moiety. The cross peak of H-1″ with C-7 (4.91/149.0 ppm) in the HMBC spectrum, as well as the cross peak of H-1″ with H-6 (4.91/ 7.30 ppm) in the NOESY spectrum demonstrated that the structure of Compound (**1**) was 8-OHDe-7-*O*-β-glucoside. The cross peaks of H-1″ with C-8 (4.93/131.9 ppm) and H-6 with C-8 (7.02/131.9 ppm) in the HMBC spectrum demonstrated that the structure of Compound (**2**) was 8-OHDe-8-*O*-β-glucoside. The key HMBC and NOESY correlations of Compounds (**1**) and (**2**) and the illustration of the biotransformation process of 8-OHDe by *BsGT110* are shown in Figure 8.

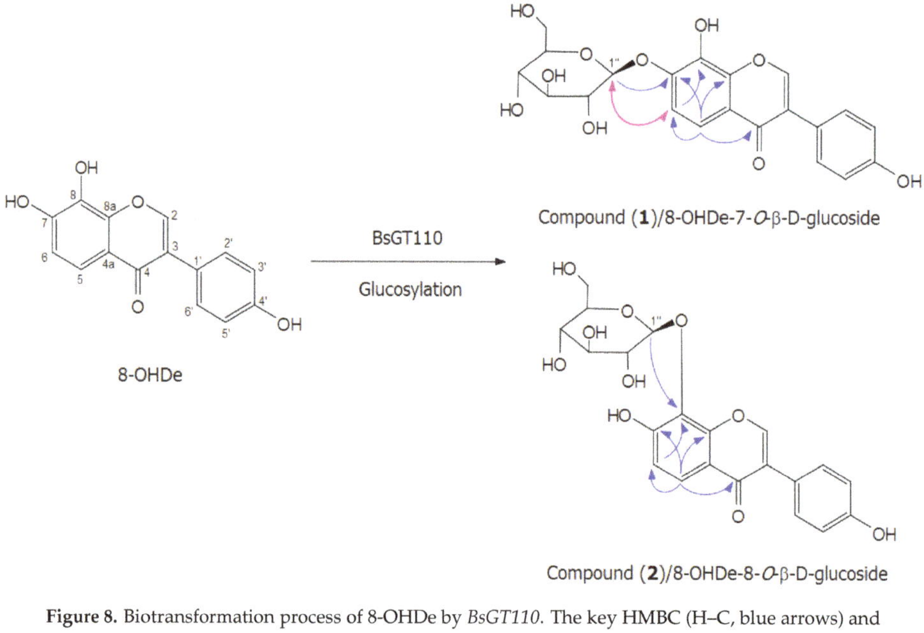

Compound (**1**)/8-OHDe-7-*O*-β-D-glucoside

BsGT110

Glucosylation

8-OHDe

Compound (**2**)/8-OHDe-8-*O*-β-D-glucoside

Figure 8. Biotransformation process of 8-OHDe by *BsGT110*. The key HMBC (H–C, blue arrows) and NOESY (pink arrows) correlations of Compound (**1**) and Compound (**2**).

2.8. Determination of Aqueous Solubility of 8-OHDe and Its Glucosides

The aqueous solubility of 8-OHDe and the two 8-OHDe glucoside derivatives was examined and is summarized in Table 1. The results revealed that the aqueous solubility of 8-OHDe-7-*O*-β-glucoside and 8-OHDe-8-*O*-β-glucoside was 9.0- and 4.9-fold, respectively, higher than that of 8-OHDe.

Table 1. Aqueous solubility of 8-OHDe and its glucoside derivatives.

Compound	Aqueous Solubility (mg/L)	Fold [1]
8-OHDe	51.3	1
8-OHDe-7-*O*-β-glucoside	462.0	9.0
8-OHDe-8-*O*-β-glucoside	251.1	4.9

[1] The fold of aqueous solubility of 8-OHDe glucoside derivatives is expressed relative to that of 8-OHDe, normalized to 1.

2.9. Determination of Stability of 8-OHDe and Its Glucosides

8-OHDe has been proven to be unstable in alkaline solution [18]. To determine the stability of 8-OHDe and its glucosides, the tested compounds were added in 50 mM of Tris at pH 8.0 and 20 °C. The residues of the compounds were monitored with UPLC at different time intervals. The results are shown in Figure 9. The amount of 8-OHDe decreased to 38% of the initial concentration after 24 h and decreased to 0.8% of the initial concentration after 96 h. In contrast, more than 90% of the two 8-OHDe glucoside derivatives remained after 96 h of incubation. The results revealed that the two 8-OHDe glucoside derivatives were much more stable than 8-OHDe in the aqueous solution.

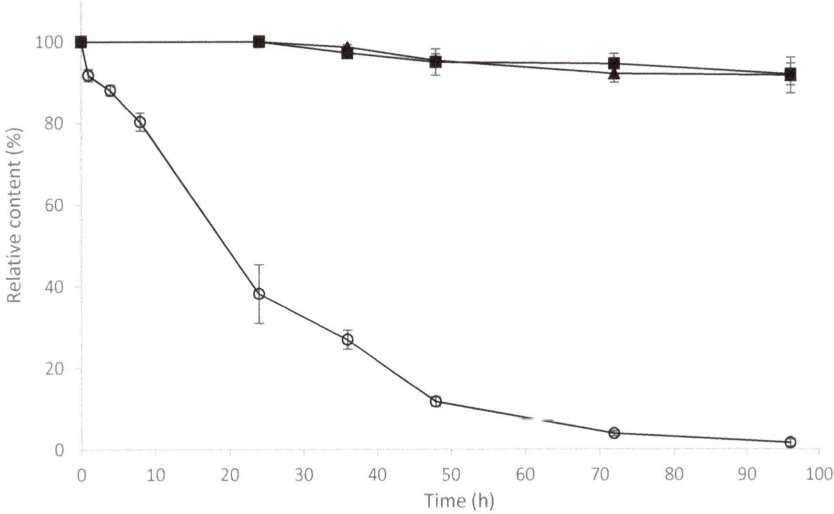

Figure 9. Stability of 8-OHDe and its glucosides. One milligram per milliliter of the tested compound was dissolved in 50 mM of Tris buffer at pH 8.0 and stored at 20 °C for 96 h. During the storage time, samples were taken out for the UPLC analysis at the determined interval times. The mean (*n* = 3) is shown, and the standard deviations are represented by error bars.

8-OHDe has been proven to possess multiple bioactivities; especially, 8-OHDe has been proven for its skin-whitening activity in vivo in the skin of mice and human volunteers [13,14]. Some skin-whitening agents maintain skin-whitening activity after glycosylation. For examples, ascorbic acid and hydroquinone are famous skin-whitening agents. Their glucoside derivatives, ascorbic acid 2-glucosdie (AA2G) and arbutin (hydroquinone-glucoside), possess potent skin-whitening activity and are the most well-known skin-whitening agents used in cosmetic markets nowadays. In our previous study, we found that *ortho*-hydroxyisoflavone glucosides possessed skin-whitening activity in vivo in the skin of mice [34]. Therefore, the 8-OHDe glucosides in the present study might have skin-whitening activity. The experiments to evaluate the in vivo skin-whitening activity in the mice skin of the 8-OHDe glucosides in the present study were conducting in our laboratory.

3. Materials and Methods

3.1. Microorganisms, Animal Cells and Chemicals

Bacillus subtilis ATCC 6633 (BCRC 10447) was purchased from the Bioresources Collection and Research Center (BCRC, Food Industry Research and Development Institute, Hsinchu, Taiwan). 8-OHDe was prepared according to Wu et al.'s [4] method. 3'-OHDe and 3'-OHGe were prepared according to our previous study [35]. 6-OHDe was purchased from Sigma (St. Louis, MO, USA). UDP-glucose was obtained from Cayman Chemical (Ann Arbor, MI, USA). All the materials needed for polymerase chain reaction (PCR), including primers, deoxyribonucleotide triphosphate and Taq DNA polymerase, were purchased from MDBio (Taipei, Taiwan). pETDuet-1 plasmid was purchased from Novagen (Madison, WI, USA). Restriction enzymes and DNA ligase were obtained from New England Biolabs (Ipswich, MA, USA). The other reagents and solvents used were of high quality and were purchased from commercially available sources.

3.2. Identification of Bacteria B. subtilis ATCC 6633 with Biotransformation Activity

Bacillus subtilis ATCC 6633 was cultivated in a 250-mL baffled Erlenmeyer flask containing 20 mL of a modified glucose-nutrient (MGN) medium (5 g/L of peptone, yeast extract, K_2HPO_4 and NaCl; 20 g/L of glucose) and 20 mg/L of 8-OHDe. After cultivation at 180 rpm, 28 °C for 24 h, 1 mL of the culture was then mixed with an equal volume of methanol. The cell debris was removed by centrifugation at $10,000 \times g$ for 10 min. The supernatant from the extracted broth was assayed with UPLC to measure the biotransformation activity.

3.3. UPLC Analysis

The UPLC system (Acquity UPLC H-Class, Waters, Milford, MA, USA) was equipped with an analytic C18 reversed-phase column (Acquity UPLC BEH C18, 1.7 μm, 2.1 i.d. × 100 mm, Waters, Milford, MA, USA). The operation conditions contained a gradient elution using water (A) containing 1% (v/v) acetic acid and methanol (B) with a linear gradient for 7 min with 35% to 80% B at a flow rate of 0.2 mL/min, an injection volume of 0.2 μL and absorbance detection at 254 nm.

3.4. Phylogenetic Analysis of BsGTs

The unrooted phylogenetic tree of the candidate genes was constructed with the maximum likelihood method, using Molecular Evolutionary Genetics Analysis (MEGA X) software (Version 10.0.4, Center for Evolutionary Functional Genomics, The Biodesign Institute, Arizona State University, Tempe, AZ, USA, 1993–2018) [32] with 500 bootstrap replications, the mtREV24+I model [31] and partial deletion.

3.5. Expression and Purification of UGT398 and UGT489

The genomic DNA of *B. subtilis* ATCC 6633 was isolated using the commercial kit Geno *Plus*™ (Viogene, Taipei, Taiwan). The three GT target genes, *BsGT110* (GenBank Protein Accession No. WP_003220110), *BsGT292* (GenBank Protein Accession No. WP_032727292) and *BsGT296* (GenBank Protein Accession No. WP_003219296), were amplified from the genomic DNA with PCR with the following primer sets: forward: 5'-CGC GAA TTC ggc taa tgt att aat gat cgg tt-3'; reverse: 5'-CGC AGA TCT tta tgc gtt ggc tga ttg agt tt-3' for *BsGT110*; forward: 5'-CGC GAA TTC gat gaa gct tgc ctt tat ctg tac ag-3'; reverse: 5'-CGC CTC GAG tta tga ttt ggc ttt cac aaa aag c-3' for *BsGT292*; forward:

5′-CGC GAA TTC Gat gaa aat agc act gat cgc cac ag-3′; reverse: 5′-CGC CTC GAG cta tct gtt ctt ctc ata cac gct g-3′ for BsGT296. Restriction enzyme-recognizing sites were designed at the forward primer (EcoRI, GAATTC) for the three BsGTs and the reverse primer (BglII, AGATCT) for *BsGT110* and (XhoI, CTCGAG) for *BsGT 292* and *BsGT296*. The amplified *BsGT* genes were subcloned into the corresponding sites of the pETDuet-1™ vector to obtain the expression vector pETDuet-BsGT (Figure 3a). In the cloning strategy, the N-terminal fusion with His-tag would allow the expressed proteins to be purified with Ni^{2+} chelate affinity chromatography. The expression vectors were transformed into *E. coli* BL21 (DE3) via electroporation to obtain the recombinant *E. coli*.

The recombinant *E. coli* was cultivated in 150 mL of Luria–Bertani (LB) medium containing 50 μg/mL of ampicillin with 200 rpm shaking at 37 °C. When the optical density at 600 nm reached 0.6, 0.2 mM IPTG was added to induce expression of the *BsGT* genes. The cells were continuously cultured in an incubator at 18 °C for another 20 h. At the end of the cultivation, the cells were harvested by centrifugation at 5000 rpm and 4 °C and washed once with 100 mL of phosphate saline buffer (PBS, 50 mM phosphate pH 6.8 and 100 mM NaCl). The cells were resuspended in 5 mL of PBS containing 25 mM of imidazole and broken by sonication at 4 °C. Cell debris was removed by centrifugation at 17,000× *g* for 30 min 4 °C. The supernatant containing the recombinant UGTs was applied on a Ni^{2+} chelate affinity column (10 i.d. × 50 mm, Ni Sepharose 6 Fast Flow, GE Healthcare, Chicago, IL, USA). After washing with 20 mL of PBS containing 25 mM of imidazole, the bound proteins were eluted with 15 mL of PBS containing 250 mM of imidazole. The elute protein was dialyzed twice against 50 mM Tris pH 8.0 and 100 mM of NaCl and then concentrated using Macrosep 10K centrifugal filters (Pall, Ann Arbor, MI, USA). The purity and molecular weights of the purified UGTs were analyzed with SDS-PAGE. The protein concentration was measured with the Bradford method with bovine serum albumin as the standard. The final purified proteins were stored at −80 °C in the presence of 50% of glycerol for use.

3.6. In Vitro Biotransformation Assay

The in vitro biotransformation was conducted with the purified BsGTs. The reaction (1 mL) containing 2 μg of the tested enzyme, 0.02 mg/mL of 8-OHDe, 0.4 mM of UDP-glucose, 10 mM of $MgCl_2$ and 50 mM of Tris at pH 8.0 was carried out at 40 °C for 30 min. After the reaction, the mixture was stopped by adding an equal volume of methanol and analyzed with UPLC. To determine the optimal condition, a standard condition was set as 2 μg of the purified *BsGT110*, 1 mg/mL of 8-OHDe, 10 mM of $MgCl_2$ and 10 mM of UDP-glucose at 50 mM of Tris at pH 8.0 and 40 °C, and the pH, temperature or metal ion in the standard condition was replaced by the tested condition. For pH testing, phosphate buffer (pH 6.0 and 7.0) and Tris buffer (pH 8.0) were used. For metal ion testing, 10 mM of either $MgCl_2$ or $CaCl_2$ were used. Relative activity was obtained by dividing the area of the summation of the two product peaks, Compound (**1**) and Compound (**2**), of the reaction in the UPLC profile by that of the reaction at the standard condition.

3.7. Scale-Up, Isolation and Identification of the Biotransformation Product

To purify the biotransformation metabolites, the reaction was scaled up to a 40-mL reaction mixture containing 20 μg of the purified *BsGT110*, 1 mg/mL of 8-OHDe, 10 mM of UDP-glucose, 10 mM of $CaCl_2$ and 50 mM of phosphate buffer at pH 7.0. After reaction at 40 °C for 30 min, 40 mL of methanol were added to stop the reaction. After being filtrated through a 0.2-μm nylon membrane, the mixture was injected into a preparative YoungLin HPLC system (YL9100, YL Instrument, Gyeonggi-do, Korea). The system was equipped with a preparative C18 reversed-phase column (Inertsil, 10 μm, 20.0 i.d. × 250 mm, ODS 3, GL Sciences, Eindhoven, The Netherlands). The operational conditions for the preparative HPLC analysis were the same as those in the UPLC analysis. The elution corresponding to the peak of the metabolite in the UPLC analysis was collected, concentrated under vacuum and then lyophilized. Finally, 17.4 mg and 24.5 mg of Compound (**1**) and Compound (**2**) were isolated, and the structures of the compounds were confirmed with NMR and mass spectral analysis. The mass analysis was performed on a Finnigan LCQ Duo mass spectrometer (ThermoQuest Corp., San Jose, CA, USA) with electrospray ionization (ESI). ^{1}H- and ^{13}C-NMR, DEPT, HSQC, HMBC, COSY and NOESY spectra were recorded on a Bruker AV-700 NMR spectrometer (Bruker Corp., Billerica, MA, USA) at ambient temperature. Standard pulse sequences and parameters were used for the NMR experiments, and all chemical shifts were reported in parts per million (ppm, δ).

3.8. Determination of Solubility

Aqueous solubility of 8-OHDe and its glucoside derivatives was examined as follows. Each compound was vortexed in d.d. H_2O for 1 h at 25 °C. The mixture was centrifuged at 10,000× g for 30 min at 25 °C and analyzed with UPLC. The concentrations of the tested compounds were determined based on their peak areas using calibration curves prepared with UPLC analyses of authentic samples.

3.9. Determination of Stability

A stock of 8-OHDe or its glucosides (100 mg/mL in dimethyl sulfoxide) was diluted 100-fold to a concentration of 1 mg/mL in 50 mM of Tris buffer at pH 8.0. Then, the diluted solutions in 1.5-mL tubes covered with alumni fossil to avoid light were placed at 20 °C for 96 h. During the storage time, samples were taken out for the UPLC analysis at the determined interval times.

4. Conclusions

8-OHDe, which has been demonstrated to possess multiple bioactivities, is a valuable isoflavone. However, the compound has very low aqueous solubility and stability, which limit its applications. In the present study, two new 8-OHDe glucoside derivatives, 8-OHDe-7-*O*-β-glucoside and 8-OHDe-8-*O*-β-glucoside, were produced through glycosylation of 8-OHDe by the glycosyltransferase *BsGT110* from *B. subtilis* ATCC 6633. The two produced 8-OHDe glucoside derivatives have higher aqueous solubility and stability than those of 8-OHDe, which could expand the use of the two new isoflavone glucosides in pharmaceutical and cosmeceutical applications in the future.

Supplementary Materials: The following are available online at http://www.mdpi.com/2073-4344/8/9/387/s1. Table S1. NMR spectroscopic data for Compound **(1)**/**(2)** (in DMSO-d6; 700MHz); Figure S1. The UPLC analysis of 24-h fermentation broth of *B. subtilis* ATCC 6633 in the absence of 8-OHDe. Figure S2. SDS-PAGE analysis of expressed and purified proteins from recombinant *E. coli* harboring pETDuet-*BsGT292* (a) and pETDuet-*BsGT292* (b). Figure S3. Biotransformation of 8-OHDe by the purified *BsGT110* (a) and *BsGT296* (b). Figure S4. The 1H-NMR (700 MHz, DMSO-d6) spectrum of Compound **(1)**. Figure S5. The 13C-NMR (176 MHz, DMSO-d6) spectrum of Compound **(1)**. Figure S6. The DEPT-90 and DEPT-135 (176 MHz, DMSO-d6) spectra of Compound **(1)**. Figure S7. The HSQC (700 MHz, DMSO-d6) spectrum of Compound **(1)**. Figure S8. The HMBC (700 MHz, DMSO-d6) spectrum of Compound **(1)**. Figure S9. The H-H COSY (700 MHz, DMSO-d6) spectrum of Compound **(1)**. Figure S10. The H-H NOESY (700 MHz, DMSO-d6) spectrum of Compound **(1)**. Figure S11. The 1H-NMR (700 MHz, DMSO-d6) spectrum of Compound **(2)**. Figure S12. The 13C-NMR (176 MHz, DMSO-d6) spectrum of Compound **(2)**. Figure S13. The DEPT-90 and DEPT-135 (176 MHz, DMSO-d6) spectra of Compound **(2)**. Figure S14. The HSQC (700 MHz, DMSO-d6) spectrum of Compound **(2)**. Figure S15. The HMBC (700 MHz, DMSO-d6) spectrum of Compound **(2)**. Figure S16. The H-H COSY (700 MHz, DMSO-d6) spectrum of Compound **(2)**. Figure S17. The H-H NOESY (700 MHz, DMSO-d6) spectrum of Compound **(2)**.

Author Contributions: Conceptualization, T.-S.C. Data curation, C.-M.C., T.-Y.W., S.-Y.Y., J.-Y.W. and T.-S.C. Methodology, C.-M.C., T.-Y.W., S.-Y.Y., J.-Y.W. and T.-S.C. Project administration, T.-S.C. Writing, original draft, C.-M.C., T.-Y.W., J.-Y.W. and T.-S.C. Writing, review and editing, C.-M.C., T.-Y.W., J.-Y.W. and T.-S.C.

Funding: This research was financially supported by grants from the National Scientific Council of Taiwan (Project No. MOST 107-2622-E-024-002-CC3).

Conflicts of Interest: The authors declare no conflicts of interest.

References

1. Franke, A.A.; Custer, L.J.; Cerna, C.M.; Narala, K.K. Quantitation of phytoestrogens in legumes by HPLC. *J. Agric. Food Chem.* **1994**, *42*, 1905–1913. [CrossRef]
2. Chang, T.S. Isolation, bioactivity, and production of *ortho*-hydroxydaidzein and *ortho*-hydroxygenistein. *Int. J. Mol. Sci.* **2014**, *15*, 5699–5716. [CrossRef] [PubMed]

3. Chang, T.S.; Ding, H.Y.; Tai, S.S.K.; Wu, C.Y. Metabolism of the soy isoflavone daidzein and genistein by the fungi used for the preparation of various fermented soybean foods. *Biosci. Biotechnol. Biochem.* **2007**, *71*, 1330–1333. [CrossRef] [PubMed]

4. Wu, S.C.; Chang, C.W.; Lin, C.W.; Hsu, Y.C. Production of 8-hydroxydaidzein polyphenol using biotransformation by *Aspergillus oryzae*. *Food Sci. Technol. Res.* **2015**, *21*, 557–562. [CrossRef]

5. Seo, M.H.; Kim, B.N.; Kim, K.R.; Lee, K.W.; Lee, C.H.; Oh, D.K. Production of 8-hydroxydaidzein from soybean extract by *Aspergillus oryzae* KACC 40247. *Biosci. Biotechnol. Biochem.* **2013**, *77*, 1245–1250. [CrossRef] [PubMed]

6. Chang, T.S.; Chao, S.Y.; Chen, Y.C. Production of *ortho*-hydroxydaidzein derivatives by a recombinant strain of *Pichia pastoris* harboring a cytochrome P450 fusion gene. *Process. Biochem.* **2013**, *48*, 426–429. [CrossRef]

7. Roh, C.; Choi, K.Y.; Pandey, B.P.; Kim, B.G. Hydroxylation of daidzein by CYP107H1 from *Bacillus subtilis* 168. *J. Mol. Catal. B Enzym.* **2009**, *59*, 248–253. [CrossRef]

8. Choi, K.Y.; Jung, E.O.; Jung, D.H.; Pandey, B.P.; Yun, H.; Park, H.Y.; Kazlauskas, R.J.; Kim, B.G. Cloning, expression and characterization of CYP102D1, a self-sufficient P450 monooxygenase from *Streptomyces avermitilis*. *FEBS J.* **2012**, *279*, 1650–1662. [CrossRef] [PubMed]

9. Funayama, S.; Anraku, Y.; Mita, A.; Komiyama, K.; Omura, S. Structure study of isoflavonoids possessing antioxidant activity from the fermentation broth of *Streptomyces sp*. *J. Antibiot.* **1989**, *42*, 1350–1355. [CrossRef] [PubMed]

10. Lo, Y.L. A potential daidzein derivative enhances cytotoxicity of epirubicin on human colon adenocarcinoma Caco-2 cells. *Int. J. Mol. Sci.* **2012**, *14*, 158–176. [CrossRef] [PubMed]

11. Chang, T.S.; Ding, H.Y.; Tai, S.S.K.; Wu, C.Y. Tyrosinase inhibitors isolated from soygerm koji fermented with *Aspergillus oryzae* BCRC 32288. *Food Chem.* **2007**, *105*, 1430–1438. [CrossRef]

12. Chang, T.S. Two potent suicide substrates of mushroom tyrosinase: 7,8,4′-trihydroxyisoflavone and 5,7,8,4′-tetrahydroxyisoflavone. *J. Agric. Food Chem.* **2007**, *55*, 2010–2015. [CrossRef] [PubMed]

13. Goh, M.J.; Park, J.S.; Bae, J.H.; Kim, D.H.; Kim, H.K.; Na, Y.J. Effects of *ortho*-dihydroxyisoflavone derivatives from Korean fermented soybean paste on melanogenesis in B16 melanoma cells and human skin equivalents. *Phytother. Res.* **2012**, *26*, 1107–1112. [CrossRef] [PubMed]

14. Tai, S.S.; Lin, C.G.; Wu, M.H.; Chang, T.S. Evaluation of depigmenting activity by 8-hydroxydaidzein in mouse B16 melanoma cells and human volunteers. *Int. J. Mol. Sci.* **2009**, *10*, 4257–4266. [CrossRef] [PubMed]

15. Fujita, T.; Funako, T.; Hayashi, H. 8-Hydroxydaidzein, an aldose reductase inhibitor from okara fermented with *Aspergillus sp*. HK-388. *Biosci. Biotechnol. Biochem.* **2004**, *68*, 1588–1590. [CrossRef] [PubMed]

16. Wu, P.S.; Ding, H.Y.; Yen, J.H.; Chen, S.F.; Lee, K.H.; Wu, M.J. Anti-inflammatory activity of 8-hydroxydaidzein in LPS-stimulated BV2 microglial cells via activation of Nrf2-antioxidant and attenuation of Akt/NF-κB-inflammatory signaling pathways, as well as inhibition of COX-2 activity. *J. Agric. Food Chem.* **2018**, *66*, 5790–5801. [CrossRef] [PubMed]

17. Kim, E.; Kang, Y.G.; Kim, J.H.; Kim, Y.J.; Lee, T.R.; Lee, J.; Kim, D.; Cho, J.Y. The antioxidant and anti-Inflammatory activities of 8-hydroxydaidzein (8-HD) in activated macrophage-Like RAW264.7 Cells. *Int. J. Mol. Sci.* **2018**, *19*, 1828. [CrossRef] [PubMed]

18. Chang, T.S. 8-Hydroxydaidzein is unstable in alkaline solutions. *J. Cosmet. Sci.* **2009**, *60*, 353–357. [CrossRef] [PubMed]

19. Shimoda, K.; Hamada, H.; Hamada, H. Synthesis of xylooligosaccharides of daidzein and their anti-oxidant and anti-allergic activities. *Int. J. Mol. Sci.* **2011**, *12*, 5616–5625. [CrossRef] [PubMed]

20. Tiwari, P.; Sangwan, R.S.; Sangwan, N.S. Plant secondary metabolism linked glycosyltransferases: An update on expanding knowledge and scopes. *Biotechnol. Adv.* **2016**, *34*, 716–739. [CrossRef] [PubMed]

21. Kim, B.G.; Yang, S.M.; Kim, S.Y.; Cha, M.N.; Ahn, J.H. Biosynthesis and production of glycosylated flavonoids in *Escherichia coli*: Current state and perspectives. *Appl. Microbiol. Biotechnol.* **2015**, *99*, 2979–2988. [CrossRef] [PubMed]

22. Hofer, B. Recent developments in the enzymatic *O*-glycosylation of flavonoids. *Appl. Microbiol. Biotechnol.* **2016**, *100*, 4269–4281. [CrossRef] [PubMed]

23. Chang, T.S.; Chiang, C.M.; Siao, Y.Y.; Wu, J.Y. Sequential Biotransformation of Antcin K by *Bacillus subtilis* ATCC 6633. *Catalysts* **2018**, *8*, 349. [CrossRef]

24. Pandey, R.P.; Gurung, R.B.; Parajuli, P.; Koirala, N.; Tuoi, L.T.; Sohng, J.K. Assessing acceptor substrates promiscuity of YjiC-mediated glycosylation towards flavonoids. *Car. Res.* **2014**, *393*, 26–31. [CrossRef] [PubMed]

25. Dai, L.; Li, J.; Yang, J.; Zhu, Y.; Men, Y.; Zeng, Y.; Cai, Y.; Dong, C.; Dai, Z.; Zhang, X.; Sun, Y. Use of a promiscuous glycosyltransferase from *Bacillus subtilis* 168 for the enzymatic synthesis of novel protopanaxtriol-type ginsenosides. *J. Agric. Food Chem.* **2017**, *66*, 943–949. [CrossRef] [PubMed]

26. Dai, L.; Li, J.; Yao, P.; Zhu, Y.; Men, Y.; Zeng, Y.; Yang, J.; Sun, Y. Exploiting the aglycon promiscuity of glycosyltransferase Bs-YjiC from *Bacillus subtilis* and its application in synthesis of glycosides. *J. Biotechnol.* **2017**, *248*, 69–76. [CrossRef] [PubMed]

27. Ko, J.H.; Kim, B.G.; Anh, J.H. Glycosylation of flavonoids with a glycosyltransferase from *Bacillus cereus*. *FEMS Mcrobiol. Lett.* **2006**, *258*, 263–268.

28. Ahn, B.C.; Kim, B.G.; Jeon, Y.M.; Lee, E.J.; Lim, Y.; Ahn, J.H. Formation of flavone di-*O*-glucosides using a glycosyltransferase from *Bacillus cereus*. *J. Microbiol. Biotechnol.* **2009**, *19*, 387–390. [CrossRef] [PubMed]

29. Zhou, M.; Hamza, A.; Zhan, C.G.; Thorson, J.S. Assessing the regioselectivity of OleD-catalyzed glcosylation with a diverse set of acceptors. *J. Nat. Prod.* **2013**, *76*, 279–286. [CrossRef] [PubMed]

30. Kim, J.H.; Kim, B.G.; Kim, J.A.; Park, Y.; Lee, Y.J.; Lim, Y.; Ahn, J.H. Glycosylation of flavonoids with *E. coli* expression glycosyltransferase from *Xanthomonas campestris*. *J. Microbiol. Biotechnol.* **2007**, *17*, 539–542. [PubMed]

31. Adachi, J.; Hasegawa, M. Model of amino acid substitution in proteins encoded by mitochondrial DNA. *J. Mol. Evol.* **1996**, *42*, 459–468. [CrossRef] [PubMed]

32. Kumar, S.; Stecher, G.; Li, M.; Knyaz, C.; Tamura, K. MEGA X: Molecular evolutionary genetics analysis across computing platforms. *Mol. Biol. Evol.* **2018**, *35*, 1547–1549. [CrossRef] [PubMed]

33. Li, C.; Ban, X.; Gu, Z.; Li, Z. Calcium ion contribution to thermostability of cyltrodetrin glycosyltransferase is closely related to calcium-binding site CaIII. *J. Agric. Food Chem.* **2013**, *61*, 8836–8841. [CrossRef] [PubMed]

34. Chang, T.S.; Wang, T.S. 3'-Isoflavone Glycosides Having Whitening and Anti-Aging Effects, Preparing Method and Use Thereof. Taiwan Patent I602580, 21 October 2017.

35. Chiang, C.M.; Wang, T.S.; Chang, T.S. Improving free radical scavenging activity of soy isoflavone glycosides daidzin and genistin by 3'-hydroxylation using recombinant *Escherichia coli*. *Molecules* **2016**, *21*, 1723. [CrossRef] [PubMed]

Article

Developing a High-Temperature Solvent-Free System for Efficient Biocatalysis of Octyl Ferulate

Shang-Ming Huang [1], Ping-Yu Wu [2], Jiann-Hwa Chen [2], Chia-Hung Kuo [3,*] and Chwen-Jen Shieh [1,*]

[1] Biotechnology Center, National Chung Hsing University, Taichung 402, Taiwan; zxzxmj2323@hotmail.com
[2] Graduate Institute of Molecular Biology, National Chung Hsing University, Taichung 402, Taiwan;
     blast2013@hotmail.com (P.-Y.W.); jhchen@dragon.nchu.edu.tw (J.-H.C.)
[3] Department of Seafood Science, National Kaohsiung University of Science and Technology,
     Kaohsiung 811, Taiwan
*   Correspondence: kuoch@nkust.edu.tw (C.-H.K.); cjshieh@nchu.edu.tw (C.-J.S.);
    Tel.: +886-7-361-7141 (ext. 23646) (C.-H.K.); +886-4-2284-0450 (ext. 5121) (C.-J.S.)

Received: 27 July 2018; Accepted: 16 August 2018; Published: 20 August 2018

Abstract: Ferulic acid esters have been suggested as a group of natural chemicals that have the function of sunscreen. The study aimed to utilize an environmentally-friendly enzymatic method through the esterification of ferulic acid with octanol, producing octyl ferulate. The Box-Behnken experimental design for response surface methodology (RSM) was performed to determine the synthesis effects of variables, including enzyme amount (1000–2000 propyl laurate units (PLU)), reaction temperature (70–90 °C), and stir speed (50–150 rpm) on the molar conversion of octyl ferulate. According to the joint test, both the enzyme amount and reaction temperature had great impacts on the molar conversion. An RSM-developed second-order polynomial equation further showed a data-fitting ability. Using ridge max analysis, the optimal parameters of the biocatalyzed reaction were: 72 h reaction time, 92.2 °C reaction temperature, 1831 PLU enzyme amount, and 92.4 rpm stir speed, respectively. Finally, the molar conversion of octyl ferulate under optimum conditions was verified to be 93.2 ± 1.5%. In conclusion, it has been suggested that a high yield of octyl ferulate should be synthesized under elevated temperature conditions with a commercial immobilized lipase. Our findings could broaden the utilization of the lipase and provide a biocatalytic approach, instead of the chemical method, for ferulic acid ester synthesis.

Keywords: ferulic acid esters; octyl ferulate; esterification; Box-Behnken design; response surface methodology; molar conversion; optimum condition

---

1. Introduction

Ferulic acid (FA) is one of the phenolic acids, which exist in natural fruits and vegetables. FA has been demonstrated as having many bioactivities, such as antioxidation [1], excellent ultraviolet-absorbing activity [2,3], and health benefits for numerous diseases, including cardiovascular-related diseases, inflammatory-related diseases, and cancers [4]. However, FA is a polar and small compound, with poor solubility in oils. Due to the limited solubility of FA in a lipophilic medium, the application of FA in cosmetics, food, and other nutraceutical industries are commonly limited due to their low dissolution rate [5,6]. Thus, it is essential to increase the solubility of FA to increase its practicality. A related study has indicated that the ester form of FA has better antioxidant activity than the original form, especially when compared to butylated hydroxytoluene (BHT) [7]. To overcome the problem of poor solubility, pharmaceutical particle technology of chemical modifications has attracted attention [8–13]. However, it is difficult to chemically synthesize such derivatives because FA is oxidation-sensitive under certain pH conditions [7,14,15]. Thus, enzymatic biocatalysis of FA as esters has been considered an alternative to chemical processes [16]. However, the environmentally-friendly synthesis process of hydrophobic derivatives, such as FA, is still a significant challenge for researchers.

The hydrophilic feruloylated derivatives have been produced by the esterification reaction of FA with monosaccharides [17]. The water-soluble derivative of glyceryl ferulate with glycerol by pectinase [18] was also synthesized through the esterification of FA. Under solvent-free conditions, the enzymatic transesterification of ethyl ferulate (EF) with triolein had a higher EF conversion of 77% [19]. In addition, hydrophobic feruloylated derivatives can be synthesized by the esterification of FA with alcohols. The strategy for esterifying hydrophilic FA with lipophilic substrates, such as fatty alcohols, can be used to modify its solubility in a lipophilic medium. In previous literature, Katsoura et al. (2009) indicated that ferulic acid esterified with different chain lengths of alkyl—such as methyl, ethyl, and octyl—and possessed a significant antioxidant capacity against lipoprotein oxidation in serum. Moreover, the study also indicated that the protective effect increased the chain length of alkyl from methyl to octyl. [20]. However, because the reaction rate of FA condensation with long chain alcoholic substrates is slow, FA might have conjugated with carboxyls and a sizable noncarboxylic region [21].

Lipases are commonly used among biocatalysts in synthetic organic chemistry [22,23]. They have been widely used to catalyze carboxylic acids and chiral alcohols, because of their great chiral recognition [24]. Among them, the lipase isoform from *Candida antarctica* (CAL-B) is used most, possessing higher enantioselectivity for a comprehensive range of substrates [25,26]. Novozym® 435 is a commercial enzyme prepared by CAL-B and immobilized on the macroporous acrylic polymer resin. Novozym® 435 is a versatile biocatalyst, mainly conducted to catalyze the hydrolysis of oils and fats, althought it can also be used to catalyze various esterification-related reactions [27]. 3D-structures of the Novozym® 435 that show a short oligopeptide helix acting as a lid and adopting different conformations as a role of detergent, provide higher mobility in its environment [28]. The performance of catalytic activity requires the flexible active site of the enzyme. Although Novozym® 435 is active in many organic solvents, in the esterification process it is needed to react under the condition of lower water activity [29]. This suggests that Novozym® 435 is stabilized mainly by hydrophobic interactions as a biocatalyst for esterification reaction [30,31]. Evidence shows that an increased rate of enzyme hydration is associated with increasing mobility of enzyme molecules [32,33]. Although Novozym® 435 was used in the synthesis of ferulic esters, the final yield remained very low (17%) after a reaction lasting several days [19]. Moreover, the study also reported that, in the ferulic esters by esterification reaction with ferulate synthesis process, the enzyme activity might be affected partly by the hydrophilicity of ferulate [34,35]. In fact, a solvent-free reaction system, which is an eco-friendly method minimizing environmental pollution, could be conducted by lipase in the synthesis system of ferulate esters. This process could reduce the reaction time but increase the synthesized yield of esters [9,23]. Additionally, the solvent-free reaction system commonly contains simple substrates and offers reaction benefits, including the maximization of substrate concentration, a higher volumetric productiveness, less environmental hazard, and a cost savings in both the reactor design of large-scale production and the chemical separation/purification of products [36]. However, to increase the yield of FA esters, it is required to develop a solvent-free system that operates at a high temperature.

As mentioned above, ferulic acid has been suggested as a natural nutraceuticals with a biological function, but because in many solvent systems FA exhibits low stability and poor solubility, its application might be limited. The present work is aimed to develop an environmentally friendly biocatalytic method, instead of chemical synthesis, for the preparation of FA esters. In this study, under solvent-free conditions, the lipase-catalyzed esterification of FA with octanol formed octyl ferulate, followed by optimizing processes that featured an experimental design using response surface methodology (RSM) to evaluate the best experimental conditions. Finally, the thermodynamic and optimum effects on a solvent-free reaction system of lipase-catalyzed biocatalysis of FA was evaluated.

2. Results and Discussion

2.1. Primary Experiment

The synthesis of octyl ferulate, catalyzed by Novozym® 435 (1500 PLU) from ferulic acid (20 mM) and octanol, was performed in a water bath at a reaction time of 72 h. The octyl ferulate catalyzed by Novozym® 435, as well as the liquid samples analyzed by high-performance liquid chromatography (HPLC), are shown in Figure 1.

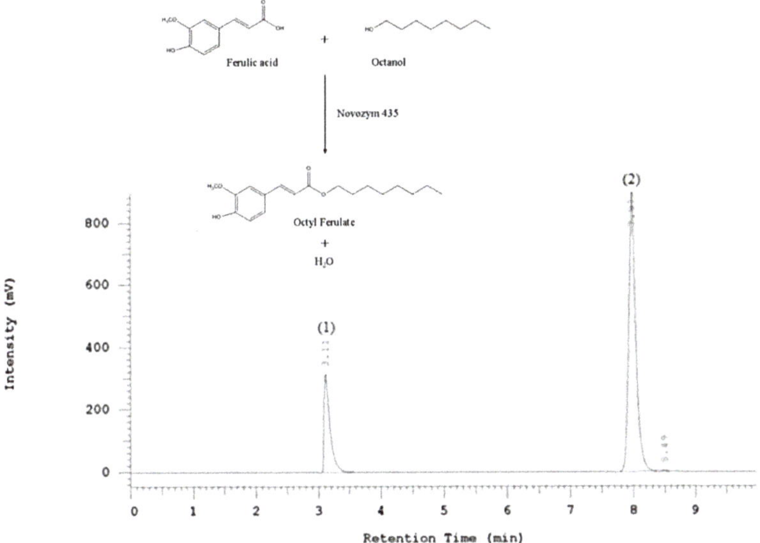

Figure 1. Scheme of octyl ferulate synthesis and high-performance liquid chromatography (HPLC) chromatogram. Peak 1 of HPLC chromatogram is ferulic acid, and peak 2 is octyl ferulate.

Moreover, the effects of reaction time and reaction temperature on molar conversion of octyl ferulate are shown in Figure 2, which illustrates an increased molar conversion observed after a reaction time of 24 h and an increased reaction temperature. Additinoally, the higher reaction temperature and increased reaction time could be observed without the weakening of enantioselectivities [37]. Although reaction temperature is a critical factor in biocatalysis, an elevated temperature may result in inactive enzymes. In previous literature, producing feruloylated lipids through the transesterification of monostearin and ethyl ferulate by catalysis has been investigated, and the optimal reaction temperature was performed at 74 °C [9]. Under a solvent-free system, a different enzymatic synthesis of feruloylated structured lipids occurred through the transesterification of ethyl ferulate with castor oil, which indicated that a high conversion of ethyl ferulate could be obtained at a reaction temperature of 90 °C [23]. The reaction temperature has a considerable influence on the equilibrium of the reversible thermodynamic reaction. Therefore, experiments performed at 70 °C, 80 °C, and 90 °C investigated the effect of temperature on biocatalysis reaction. As illustrated in Figure 2, the molar conversion increased with an elevating reaction temperature. At a reaction temperature of 90 °C, molar conversion reached a high value of 88.4%; the fastest reaction rate (observed by the slope of the linear curves) was also determined. When the reaction time was over 72 h, the time course was leveled-off. The reason for this may be that the viscosity of the reaction mixture gradually increased over reaction time, leading to a decrease in mutual solubility and inhibiting the diffusion of substrates [38]. Thus, reducing mass transfers and unfavorable interactions between substrates and enzyme particles might occur in the catalysis process. However, a longer reaction time could lead to a higher molar conversion. Thus, the following experimental design was conducted at a reaction time of 72 h.

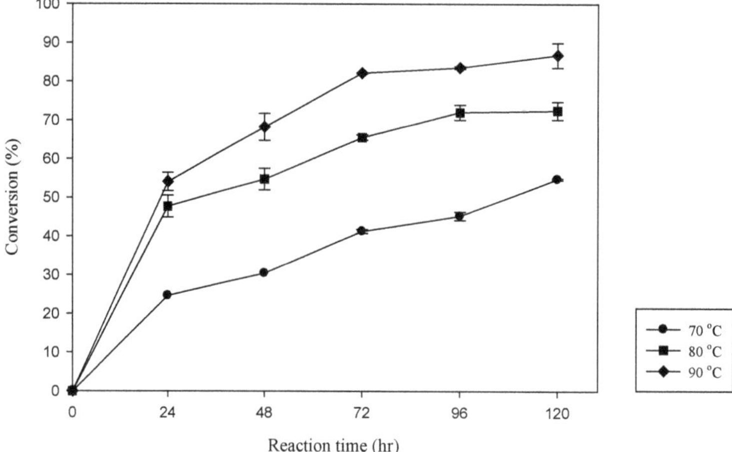

Figure 2. Effects of reaction temperature and time on Novozym® 435-catalyzed synthesis of octyl ferulate. The reaction was employed in a water bath at a reaction temperature of 70–90 °C and enzyme amount of 1500 propyl laurate units (PLU).

2.2. Model Fitting

In our preliminary study, we explored the relationship between the influence factors and synthesis yield of octyl ferulate catalyzed by Novozym® 435, testing various independent variables, such as reaction time, reaction temperature, the molar ratio of octanol and ferulic acid, enzyme amount, stirring speed, and so on. Moreover, our preliminary data indicated that the reaction temperature, enzyme amount, and stirring speed were synthesis-dependent variables in response to Novozym® 435-catalysed octyl ferulate. Thus, we chose these factors to investigate the optimal process of octyl synthesis by Box-Behnken design (BBD). In order to systematically realize the interactions between enzyme amount, reaction temperature, and stir speed of the octyl ferulate synthesis, a three-level, three-factor BBD was used to test 15 experiments (treatments). Using statistical response surface methodology (RSM) modelling, we determined the experimental data for process optimization of octyl ferulate production. This study aimed to develop and evaluate a statistical approach in order to thoroughly realize the relationship between experimental variables and lipase-catalyzed responses. Thus, the process could be fully optimized before the scaling-up production, which decreased the cost and time to obtain a high-quality product. Compared with the single-factor-at-a-time design in most studies, our study's combination of RSM and BBD was more effective in reducing experimental runs investigating the optimized process of octyl ferulate biocatalysis.

The experimental conditions and results of a three-level, three-factor BBD are shown in Table 1. The response surface regression (RSREG) process for SAS was employed to fit the second-order polynomial Equation of the experimental molar conversions. Among these treatments, the highest molar conversion (88.4 ± 3.1%) was treatment 8 (reaction temperature 90 °C, enzyme amount 1500 PLU and stirring speed 150 rpm); the lowest molar conversion (42.4 ± 5.7%) was treatment 1 (reaction temperature of 70 °C, enzyme amount of 1000 PLU and stirring speed of 100 rpm). From the SAS output of RSREG, the second-order polynomial Equation (1) is given below:

$$Y = -299.1 + 5.6825X_1 + 0.06625X_2 + 0.22X_3 - 0.00031X_1X_2$$
$$-0.0035X_1X_3 - 0.00005X_2X_3 - 0.02025X_1{}^2 \tag{1}$$
$$-0.0000059X_2{}^2 + 0.00091X_3{}^2$$

Table 1. Box-Behnken design experiments and observed data of molar conversion.

Treatment No. [a]	Experimental Factors [b]			Molar Conversion [c] (%)
	X_1 (°C)	X_2 (PLU)	X_3 (rpm)	
1	−1(70)	−1(1000)	0(100)	42.4 ± 5.7
2	1(90)	−1(1000)	0(100)	75.2 ± 0.7
3	−1(70)	1(2000)	0(100)	61.7 ± 3.6
4	1(90)	1(2000)	0(100)	88.3 ± 0.5
5	−1(70)	0(1500)	−1(50)	49.4 ± 1.1
6	1(90)	0(1500)	−1(50)	88.3 ± 0.1
7	−1(70)	0(1500)	1(150)	56.5 ± 0.7
8	1(90)	0(1500)	1(150)	88.4 ± 3.1
9	0(80)	−1(1000)	−1(50)	56.4 ± 0.1
10	0(80)	1(2000)	−1(50)	80.2 ± 3.8
11	0(80)	−1(1000)	1(150)	64.7 ± 4.9
12	0(80)	1(2000)	1(150)	83.5 ± 1.1
13	0(80)	0(1500)	0(100)	71.1 ± 1.4
14	0(80)	0(1500)	0(100)	70.2 ± 2.8
15	0(80)	0(1500)	0(100)	70.2 ± 3.5

[a] The treatments were employed in a random order. [b] X_1: reaction temperature; X_2: enzyme amount; X_3: stir speed. [c] Molar conversion for octyl ferulate shows means ± SD of duplicated experiments.

Furthermore, the analysis of variance (ANOVA) indicated that the second-order polynomial equation had a significant correlation between the experimental response (molar conversion) and dependent variables, which had a very small p-value of 0.0001 and a satisfactory coefficient of $R^2 = 0.9887$. Additionally, the total effects of these three experimental variables on the molar conversion were investigated by a joint test. The results indicated that the reaction temperature (X_1) and enzyme amount (X_2) were the most critical factors, statistically showing a significant effect ($p < 0.001$) on the responded molar conversion. The stirring speed (X_3) was statistically insignificant on the responded molar conversion.

2.3. Optimal Synthesis Conditions

The optimal biocatalysis of octyl ferulate was further determined by the ridge max analysis, indicating that the highest molar conversion was 91.74 ± 2.3% at 72 h, temperature 92.2 °C, enzyme amount 1831 PLU, and stir speed 92.4 rpm (Figure 3). Figure 4 indicates the correlation between the predicted and experimental values of the average cutting speed in the RSM model. A certification experiment, performed in the investigated optimal conditions, could obtain a 93.2 ± 1.5% molar conversion, which was similar to the RSM-predicted molar conversion, thus indicating that the predicted model in this study was successfully established. Thus, the lipase-catalyzed biocatalysis of octyl ferulate could be carried out with an easy method.

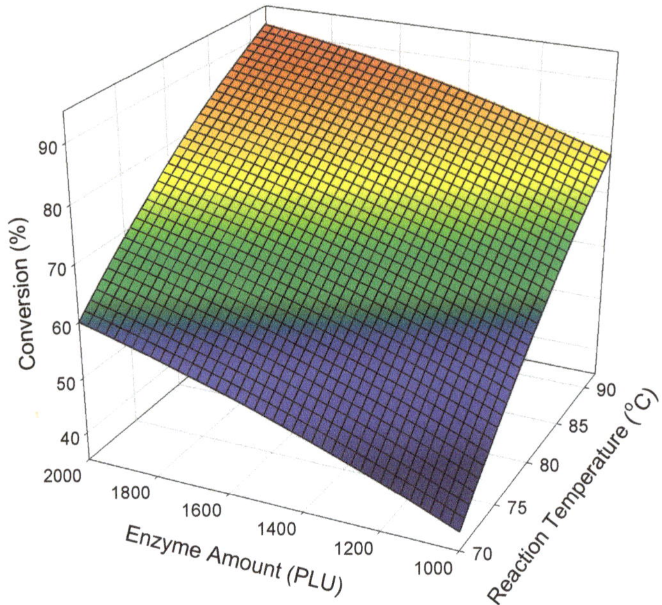

Figure 3. Response surface plots show the correlation between the molar conversion of octyl ferulate synthesis and reaction parameters (enzyme amount and reaction temperature).

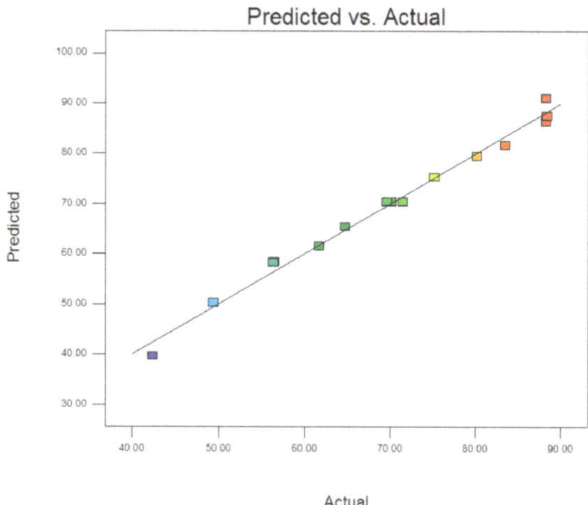

Figure 4. Comparison of the experimental data with those predicted by the response surface methodology (RSM) model.

Additionally, to confirm the enzyme reusability, the immobilized lipase used for octyl ferulate synthesis was determined under optimum conditions. Novozym® 435 was recovered from the reaction medium after synthesis, followed by a direct reuse in the next batch. When the immobilized lipase was reused five times, the molar conversion of octyl ferulate remained higher than 90% with very little loss of enzymatic activity (Figure 5).

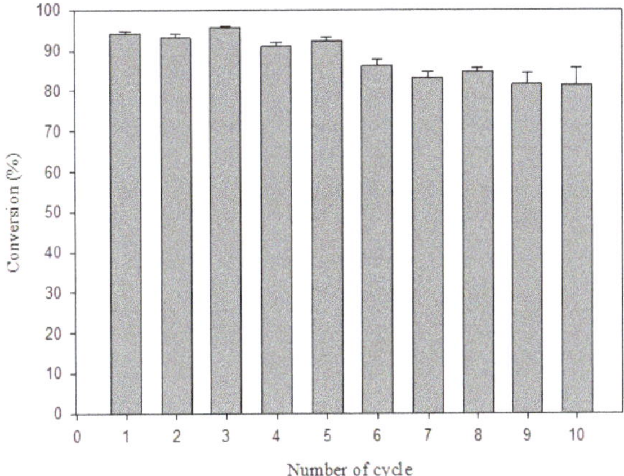

Figure 5. Reusable cycles of Novozym® 435 in the synthetic process of octyl ferulate under indicated optimal condition.

The data indicated that the immobilized lipase could keep enzymatic stability under the conditions of long-term octanol exposure and high reaction temperature (90 °C). Therefore, the data confirmed that the Novozym® 435 could be efficiently used for the synthesis of octyl ferulate and that the enzymatic stability was enough to reuse.

3. Materials and Methods

3.1. Materials

A total of 10,000 U/g (propyl laurate units, PLU) of immobilized lipase Novozym® 435 from *Candida antarctica* B (EC3.1.1.3) (on a macroporous acrylic resin) was purchased from Novo Nordisk Bioindustrials Inc. (Copenhagen, Denmark). Ferulic acid (FA), acetic acid, 2-methyl-2-butanol, methanol, hexanol, and octanol were obtained from Sigma Chemical Co. (St. Louis, MO, USA). A 4-Å molecular sieve was purchased from Davison Chemical (Baltimore, MD, USA). All chemicals and reagents were of analytical grade.

3.2. Enzymatic Synthesis of Octyl Ferulate

In this study, the synthesis process was performed under a solvent-free system in order to biocatalyze ferulic esters, using the maximum concentration of ferulic acid, which helped its participation in esterification. However, the best solubility of ferulic acid in octanol was also determined by a maximum concentration of 20 mM. Therefore, 20 mM of ferulic acid was employed in this study. All materials were dehydrated using molecular sieve (4 Å) for 24 h before use. The 20 mM of ferulic acid and Novozym® 435 were thoroughly mixed with octanol (1 mL) in sealed dark vials, and reacted in a water bath for 72 h, under different experimental conditions of temperature, enzyme amount, and stir speed, as shown in Table 1. The reacted samples were further analyzed by high-performance liquid chromatography (HPLC). Samples were centrifuged and diluted with hexanol/2-methyl-2-butanol (1:100). Analyses were done according to the procedure modified by Huang et al. [36]. The flow rate was set at 1.0 mL/min, followed by detection at a wavelength of 325 nm. Molar conversion was calculated based on the peak areas of the sample.

3.3. Experimental Design

In this study, a three-level, three-factor Box-Behnken design requiring 15 experiments was employed. To avoid bias, the 15 runs were done in a random order. The variables and their response levels selected for the synthesis of octyl ferulate were: Reaction temperature (70–90 °C), enzyme amount (1000–2000 PLU), and stir speed (50–150 rpm). All of the experiments were performed at a reaction time of 72 h. Table 1 shows the independent factors (X_1), levels, and experimental design of the coded and un-coded values.

3.4. Statistical Analysis

All data were analyzed by the procedure of response surface regression (RSREG) with SAS software to fill in the second-order polynomial equation, as shown in Equation (2):

$$Y = \beta_0 + \sum_{i=1}^{3} \beta_i x_i + \sum_{i=1}^{3} \beta_{ii} x_i^2 + \sum_{i=1}^{2} \sum_{j=i+1}^{3} \beta_{ij} x_i x_j \tag{2}$$

Y is the molar conversion of octyl ferulate; β_0 represents a constant; β_i, β_{ii}, and β_{ij} represent constant coefficients; and xi and xj are uncoded variables. The suffixes i and j are shown in Equation (1) with the three variables representing: x_1 for reaction temperature, x_2 for enzyme amount, and x_3 for stir speed. The decision of ridge max in the SAS software was used to calculate the estimation of maximum response ridge, in which the radius increased from the center of the original design.

4. Conclusions

In this study, the esterification of ferulate with octanol catalyzed by lipase in a solvent-free system was well investigated. The immobilized Novozym® 435 could be used to synthesize octyl ferulate. Both three-level, three-factor BBD and RSM were employed successfully for the experimental design. An environmentally friendly experimental model for the octyl ferulate synthesis was built, and the optimal conditions for biocatalysis had a reaction time of 72 h, a reaction temperature of 92.2 °C, an enzyme amount of 1831 PLU, and a stir speed of 92.4 rpm. In order to obtain a molar conversion of 93.2 ± 1.5%, we employed a synthesis of octyl ferulate by lipase biocatalysis under the optimal condition. Notably, our results indicated that a high reaction temperature significantly affected the efficiency of a lipase-catalyzed ester synthesis under a solvent-free system. Overall, as compared to the chemical methods, our present study suggests that the lipase-catalyzed and environmentally-friendly synthesis of esters, when under a solvent-free reaction system at a high reaction temperature and without a marked loss of enzymatic activity, could offer a significant reference for an industrial preparation process.

Author Contributions: C.-J.S., J.-H.C. and C.-H.K. conceived and designed the experiments; P.-Y.W. performed the experiments; S.-M.H. and P.-Y.W. analyzed the data; S.-M.H. and C.-H.K. wrote and revised the paper.

Acknowledgments: The authors are very appreciative of all the support from the Ministry of Science and Technology under the grant number of MOST-101-2313-B-005-040-MY3.

Conflicts of Interest: There is no any conflict of interest regarding the publication of this article.

References

1. Liu, J.; Wen, X.Y.; Lu, J.F.; Kan, J.; Jin, C.H. Free radical mediated grafting of chitosan with caffeic and ferulicacids: Structures and antioxidant activity. *Int. J. Biol. Macromol.* **2014**, *65*, 97–106. [CrossRef] [PubMed]
2. Choo, W.S.; Birch, E.J. Radical scavenging activity of lipophilized products from lipase-catalyzed transesterification of triolein with cinnamic and ferulic acids. *Lipids* **2009**, *44*, 145–152. [CrossRef] [PubMed]
3. Itagaki, S.; Kurokawa, T.; Nakata, C.; Saito, Y.; Oikawa, S.; Kobayashi, M.; Hirano, T.; Iseki, K. In vitro and in vivo antioxidant properties of ferulic acid: A comparative study with other natural oxidation inhibitors. *Food Chem.* **2009**, *114*, 466–471. [CrossRef]

4. Mancuso, C.; Santangelo, R. Ferulic acid: Pharmacological and toxicological aspects. *Food Chem. Toxicol.* **2014**, *65*, 185–195. [CrossRef] [PubMed]

5. Karboune, S.; St-Louis, R.; Kermasha, S. Enzymatic synthesis of structured phenolic lipids by acidolysis of flaxseed oil with selected phenolic acids. *J. Molec. Catal. B Enzym.* **2008**, *52–53*, 96–105. [CrossRef]

6. Zheng, Y.; Branford-White, C.; Wu, X.M.; Wu, C.Y.; Xie, J.G.; Quan, J.; Zhu, L.M. Enzymatic synthesis of novel feruloylated lipids and their evaluation as antioxidants. *J. Amer. Oil Chem. Soc.* **2010**, *87*, 305–311. [CrossRef]

7. Kikuzaki, H.; Hisamoto, M.; Hirose, K.; Akiyama, K.; Taniguchi, H. Antioxidant properties of ferulic acid and its related compounds. *J. Agric. Food Chem.* **2002**, *50*, 2161–2168. [CrossRef] [PubMed]

8. Reddy, K.K.; Ravinder, T.; Kanjilal, S. Synthesis and evaluation of antioxidant and antifungal activities of novel ricinoleate-based lipoconjugates of phenolic acids. *Food Chem.* **2012**, *134*, 2201–2207. [CrossRef] [PubMed]

9. Sun, S.D.; Song, F.F.; Bi, Y.L.; Yang, G.L.; Liu, W. Solvent-free enzymatic transesterification of ethyl ferulate and monostearin: Optimized by response surface methodology. *J. Biotechnol.* **2012**, *164*, 340–345. [CrossRef] [PubMed]

10. Xin, J.; Zhang, L.; Chen, L.; Zheng, Y.; Wu, X.; Xia, C. Lipase-catalyzed synthesis of ferulyl oleins in solvent-free medium. *Food Chem.* **2009**, *112*, 640–645. [CrossRef]

11. Yang, Z.; Guo, Z.; Xu, X. Enzymatic lipophilisation of phenolic acids through esterification with fatty alcohols in organic solvents. *Food Chem.* **2012**, *132*, 1311–1315. [CrossRef] [PubMed]

12. Yang, H.; Mu, Y.; Chen, H.; Xiu, Z.; Yang, T. Enzymatic synthesis of feruloylated lysophospholipid in a selected organic solvent medium. *Food Chem.* **2013**, *141*, 3317–3322. [CrossRef] [PubMed]

13. Fang, X.; Shima, M.; Kadota, M.; Tsuno, T.; Adachi, S. Suppressive effect of alkyl ferulate on the oxidation of linoleic acid. *Biosci. Biotechnol. Biochem.* **2006**, *70*, 457–461. [CrossRef] [PubMed]

14. Yang, X.-Z.; Diao, X.-J.; Yang, W.-H.; Li, F.; He, G.-W.; Gong, G.-Q.; Xu, Y.-G. Design, synthesis and antithrombotic evaluation of novel dabigatran prodrugs containing methyl ferulate. *Bioorg. Med. Chem. Lett.* **2013**, *23*, 2089–2092. [CrossRef] [PubMed]

15. Giuliani, S.; Piana, C.; Setti, L.; Hochkoeppler, A.; Pifferi, P.G.; Williamson, G.; Faulds, C.B. Synthesis of pentylferulate by a feruloyl esterase from Aspergillus niger using water-in-oil microemulsions. *Biotechnol. Lett.* **2001**, *23*, 325–330. [CrossRef]

16. Saija, A.; Tomaino, A.; Cascio, R.L.; Trombetta, D.; Proteggente, A.; Pasquale, A.D.; Uccella, N.; Bonina, F. Ferulic and caffeic acids as potential protective agents against photooxidative skin damage. *J. Sci. Food Agric.* **1999**, *79*, 476–480. [CrossRef]

17. Stamatis, H.; Sereti, V.; Kolisis, F.N. Enzymatic synthesis of hydrophilic and hydrophobic derivatives of natural phenolic acids in organic media. *J. Mol. Catal. B Enzym.* **2001**, *11*, 323–328. [CrossRef]

18. Matsuo, T.; Kobayashi, T.; Kimura, Y.; Tsuchiyama, M.; Oh, T.; Sakamoto, T.; Adachi, S. Synthesis of glyceryl ferulate by immobilized ferulic acid esterase. *Biotechnol. Lett.* **2008**, *30*, 2151–2156. [CrossRef] [PubMed]

19. Compton, D.L.; Laszlo, J.A.; Berhow, M.A. Lipase-catalyzed synthesis of ferulate esters. *J. Am. Oil Chem. Soc.* **2000**, *77*, 513–519. [CrossRef]

20. Katsoura, M.H.; Polydera, A.C.; Tsironis, L.D.; Petraki, M.P.; Rajačić, S.K.; Tselepis, A.D.; Stamatis, H. Efficient enzymatic preparation of hydroxycinnamates in ionic liquids enhances their antioxidant effect on lipoproteins oxidative modification. *New Biotechnol.* **2009**, *26*, 83–91. [CrossRef] [PubMed]

21. Kobayashi, T.; Adachi, S.; Matsuno, R. Lipase-catalyzed condensation of *p*-methoxyphenethyl alcohol and carboxylic acids with different steric and electrical properties in acetonitrile. *Biotechnol. Lett.* **2003**, *25*, 3–7. [CrossRef] [PubMed]

22. Zagozda, M.; Plenkiewicz, J. Biotransformations of 2,3-epoxy-3-arylpropanenitriles by *Debaryomyces hansenii* and *Mortierella isabellina* cells. *Tetrahedron Asymmetry* **2008**, *19*, 1454–1460. [CrossRef]

23. Sun, S.; Zhu, S.; Bi, Y. Solvent-free enzymatic synthesis of feruloylated structured lipids by the transesterification of ethyl ferulate with castor oil. *Food Chem.* **2014**, *158*, 292–295. [CrossRef] [PubMed]

24. Bornscheuer, U.T.; Kazlauskas, R.J. *Hydrolases in Organic Synthesis*; Wiley-VCH: Weinheim, Germany, 2006; pp. 61–183.

25. Xu, D.; Li, Z.; Ma, S. Novozym-435-catalyzed enzymatic separation of racemic propargylic alcohols. A facile route to optically active terminal aryl propargylic alcohols. *Tetrahedron Lett.* **2003**, *44*, 6343–6346. [CrossRef]

26. Jacobsen, E.E.; van Hellemond, E.; Moen, A.R.; Prado, L.C.V.; Anthonsen, T. Enhanced selectivity in Novozym 435 catalyzed kinetic resolution of secondary alcohols and butanoates caused by the (*R*)-alcohols. *Tetrahedron Lett.* **2003**, *44*, 8453–8455. [CrossRef]

27. Rajendran, A.; Palanisamy, A.; Thangavelu, V. Lipase catalyzed ester synthesis for food processing industries. *Braz. Arch. Biol. Technol.* **2009**, *52*, 207–219. [CrossRef]

28. Uppenberg, J.; Hansen, M.T.; Patkar, S.; Jones, T.A. The sequence, crystal structure determination and refinement of two crystal forms of lipase B from *Candida Antarctica. Structure* **1994**, *2*, 293–308. [CrossRef]

29. Jaeger, K.-E.; Eggert, T. Lipases for biotechnology. *Curr. Opin. Biotechnol.* **2002**, *13*, 390–397. [CrossRef]

30. Mei, Y.; Miller, L.; Gao, W.; Gross, R.A. Imaging the distribution and secondary structure of immobilized enzymes using infrared microspectroscopy. *Biomacromolecules* **2003**, *4*, 70–74. [CrossRef] [PubMed]

31. Cabrera, Z.; Fernandez-Lorente, G.; Fernandez-Lafuente, R.; Palomo, J.M.; Guisan, J.M. Novozym 435 displays very different selectivity compared to lipase from *Candida Antarctica* B adsorbed on other hydrophobic supports. *J. Mol. Catal. B Enzym.* **2009**, *57*, 171–176. [CrossRef]

32. Burke, P.A.; Griffin, R.G.; Klibanov, A.M. Solid-state nuclear magnetic resonance investigation of solvent dependence of tyrosyl ring motion in an enzyme. *Biotechnol. Bioeng.* **1993**, *42*, 87–94. [CrossRef] [PubMed]

33. Affleck, R.; Xu, Z.F.; Suzawa, V.; Focht, K.; Clark, D.S.; Dordick, J.S. Enzymatic catalysis and dynamics in low-water environments. *Proc. Natl. Acad. Sci. USA* **1992**, *89*, 1100–1104. [CrossRef] [PubMed]

34. Iwai, M.; Okumura, S.; Tsujisaka, Y. Synthesis of terpene alcohol esters by lipase. *Agric. Biol. Chem.* **1980**, *44*, 2731–2732.

35. Langrand, G.; Rondot, N.; Triantaphylides, C.; Baratti, J. Short chain flavour esters synthesis by microbial lipases. *Biotechnol. Lett.* **1990**, *12*, 581–586. [CrossRef]

36. Huang, K.-C.; Li, Y.; Kuo, C.-H.; Twu, Y.-K.; Shieh, C.-J. Highly efficient Synthesis of an emerging lipophilic antioxidant: 2-ethylhexyl ferulate. *Molecules* **2016**, *21*, 478. [CrossRef] [PubMed]

37. Bezerra, M.A.; Santelli, R.E.; Oliveira, E.P.; Villar, L.S.; Escaleira, L.A. Response surface methodology (RSM) as a tool for optimization in analytical chemistry. *Talanta* **2008**, *76*, 965–977. [CrossRef] [PubMed]

38. Jeon, N.Y.; Ko, S.-J.; Won, K.; Kang, H.-Y.; Kim, B.T.; Lee, Y.S.; Lee, H. Synthesis of alkyl (*R*)-lactates and alkyl (*S*,*S*)-*O*-lactyllactates by alcoholysis of *rac*-lactide using Novozym 435. *Tetrahedron Lett.* **2006**, *47*, 6517–6520. [CrossRef]

Article

Understanding (R) Specific Carbonyl Reductase from *Candida parapsilosis* ATCC 7330 [CpCR]: Substrate Scope, Kinetic Studies and the Role of Zinc

Vinay Kumar Karanam [1], Debayan Chaudhury [1] and Anju Chadha [1,2,*

[1] Laboratory of Bioorganic Chemistry, Department of Biotechnology, Indian Institute of Technology Madras, Chennai 600036, India
[2] National Centre for Catalysis Research, Indian Institute of Technology Madras, Chennai 600036, India
* Correspondence: anjuc@iitm.ac.in; Tel.: +91-044-2257-4106

Received: 26 July 2019; Accepted: 18 August 2019; Published: 21 August 2019

Abstract: CpCR, an (R) specific carbonyl reductase, so named because it gave (R)-alcohols on asymmetric reduction of ketones and ketoesters, is a recombinantly expressed enzyme from *Candida parapsilosis* ATCC 7330. It turns out to be a better aldehyde reductase and catalyses cofactor (NADPH) specific reduction of aliphatic and aromatic aldehydes. Kinetics studies against benzaldehyde and 2,4-dichlorobenzaldehyde show that the enzyme affinity and rate of reaction change significantly upon substitution on the benzene ring of benzaldehyde. CpCR, an MDR (medium chain reductase/dehydrogenase) containing both structural and catalytic Zn atoms, exists as a dimer, unlike the (S) specific reductase (SRED) from the same yeast which can exist in both dimeric and tetrameric forms. Divalent metal salts inhibit the enzyme even at nanomolar concentrations. EDTA chelation decreases CpCR activity. However, chelation done after the enzyme is pre-incubated with the NADPH retains most of the activity implying that Zn removal is largely prevented by the formation of the enzyme-cofactor complex.

Keywords: MDR—medium-chain reductase/dehydrogenase; ADH—alcohol dehydrogenase; enzyme kinetics; EDTA (Ethylenediaminetetraacetic acid) chelation; ultrafiltration

---

1. Introduction

The MDR superfamily is a part of the oxidoreductase class and contains a family of zinc-dependent alcohol dehydrogenases [1]. MDRs are hypothesized to have evolved from SDR (short-chain reductases/dehydrogenases) superfamily and later incorporated zinc atoms within themselves to facilitate divergence in catalytic abilities [2]. CpCR belongs to MDR superfamily and is reported to catalyse reductions of ketoesters, ketones and aldehydes leading to the production of some important pharmaceutical precursors [3]. It is one of the important enzymes present in *Candida parapsilosis* ATCC 7330, which is a well-known whole-cell biocatalyst [4]. CpCR, a heterodimer (PDB: 4OAQ), has two different Zn atoms viz. catalytic Zn and the structural Zn. The former is coordinated to two Cys, His and a water while the latter is coordinated to four Cys residues and lies away from the active site. Aldehyde reduction by various ADHs from horse liver, human liver and *Saccharomyces* sp. is well established [5–7]. CpCR reduces aliphatic and aromatic aldehydes with higher activity compared to other carbonyl substrates.

Even though a lot of literature on understanding the role of Zn in MDR superfamily exists [8–17], still there is some ambiguity in the function of structural Zn [8,9,16,17]. Chelation studies with multi-dentate ligands, like EDTA and 1,10-phenanthroline on ADHs, indicate that they significantly affect the activity by chelating one of the Zn atoms [9,17,18]. Dithiothreitol (DTT) at higher concentrations is known to cause heat lability of yeast ADH (YADH) by changing the Zn stoichiometry in the enzyme [9]. Cofactor binding to the liver ADH (LADH) induces a large conformational change where the two domains (catalytic and cofactor binding domains) rotate around 10 degrees to close the active site cleft [19]. A similar observation was made in alcohol dehydrogenase from *Arabidopsis thaliana* but the mechanism is different from that of LADH [20]. It is also established that the conformational change induced by cofactor binding requires the presence of the nicotinamide part of NAD(P)H, while the binding of ADP-ribose does not induce such a change [21]. Recently, cofactor binding to various ADHs was studied using circular dichroism wherein the orientation of nicotinamide ring of the cofactor at the active site could be observed [22]. Another study on cofactor binding shows that NAD^+ and NADH adopt different structures in water, but both fit in the enzyme's active site in a semi-extended conformation [23]. These studies are essential in understanding the initial step (binding of the cofactor to the active site) of the reactions catalysed by NAD(P)H-dependent ADHs. Cofactor switching is also an important aspect in obtaining enzymes with better catalytic ability and applications in metabolic engineering [24–27]. However, the effect of this changed configuration upon cofactor binding on enzyme activity has not been probed systematically to date. This is of importance because in nature most enzymes exist bound to their natural cofactor as evidenced by typically low K_d values of the cofactor [22].

In this study we used various concentrations of EDTA for chelation studies against CpCR and employed ultrafiltration for rapid removal of EDTA. To the best of our knowledge this is the first report to elucidate the kinetic characteristics of cofactor-enzyme complex.

2. Results and Discussion

2.1. Purification of CpCR, Expanding Its Substrate Scope and Kinetic Studies

The purification protocol was modified, keeping in mind the yield and the stability of the enzyme. Compared to the previous protocol [28], the modified protocol increased the yield ten times and fold purification by 3.7 times. CpCR, an MDR, is a Zn-containing enzyme and it is very important that the Zn coordination stays unaffected by the buffer conditions in which it is purified/stored. Earlier purifications of CpCR had DTT in the buffers to maintain a reducing environment for the four free cysteine residues present in the enzyme. DTT, a reducing agent is known to reduce the Cys residues coordinated to the zinc and release it [9]. CpCR and YADH belong to the same MDR superfamily. Thus, the enzyme was purified without the addition of DTT in the buffering system and the effects were clear with the increase in the activity by more than three times. The presence of $MgCl_2$ in the storage buffer was also omitted as the Mg ion does not have any significant interactions with the protein surface (PDB: 4OAQ). HEPES replacing Tris-HCl buffer was based on the fact that the pH of the Tris buffer is sensitive to changes in temperature.

Our previous study reported an asymmetric reduction of ketones and ketoesters by CpCR, but the activity with aldehydes was better [3]. Thus, in this work, a detailed study of aldehydes, i.e., aliphatic and various substituents of benzaldehyde as the substrates for CpCR, was carried out. Aldehyde reduction is NADPH-specific. Aliphatic aldehydes show 60–70% activity as that of benzaldehyde (Table 1). Any substitution on any position of benzaldehyde decreases the activity, due to electronic and steric effects. 2, 4-dichlorobenzaldehyde (substrate 15) shows the least activity. Among the *ortho* and *para* substituted benzaldehydes, electron withdrawing groups like NO_2 and CN (substrates 13 & 12) show less activity as they can destabilise the benzene ring. Bromo and floro substitutions along with electron donating groups such as CH_3 and $O-CH_3$ at *ortho* and *para* positions (substrates 4, 9, 10 and 11) show comparatively better activity than substrates which contain electron-withdrawing groups (substrates 6, 12 and 13). It is expected that substitutions on *ortho* and *para* positions behave similarly but in the case of *ortho* bromo (substrate 5), steric effects dominate. Substrate 9 shows comparatively better activity than substrates 10 and 11 due to the presence of the smaller methyl group.

Table 1. The specific activity of CpCR against various aldehydes that were not reported earlier.

Entry	Substrate	Specific Activity (U mg^{-1}) [1]
1	Propanal	24.98 ± 1.06
2	Pentanal	19.29 ± 0.59
3	Hexanal	21.9 ± 0.64
4	o-Fluorobenzaldehyde	23.06 ± 0.94
5	o-Bromobenzaldehyde	6.76 ± 0.63
6	o-Cyanobenzaldehyde	21.17 ± 0.73
7	m-Fluorobenzaldehyde	6.53 ± 0.82
8	m-Bromobenzaldehyde	20.79 ± 0.76
9	p-Methylbenzaldehyde	26.46 ± 0.92
10	p-Bromobenzaldehyde	21.0 ± 0.53
11	p-Methoxybenzaldehyde	19.69 ± 0.97
12	p-Cyanobenzaldehyde	18.63 ± 0.72
13	p-Nitrobenzaldehyde	13.35 ± 0.69
14	Benzaldehyde	33.96 ± 1.02 [2]
15	2, 4-dichlorobenzaldehyde	3.8 ± 0.64

[1] One unit of the enzyme activity is defined as the amount of enzyme that oxidizes 1 μmol of NADPH per minute at 30 °C. [2] Previously reported [3]. Put here for comparative analysis with other substrates.

The kinetic parameters, i.e., enzyme affinity and the catalytic rates of CpCR with two different substrates (benzaldehyde and 2, 4-dichlorobenzaldehyde) show drastic differences in activity (Table 2).

The dramatic decrease in activity with 2,4-dichlorobenzaldehyde could be because of its poor fit in the substrate cleft as reflected in >20 fold higher K_m value as compared to that of benzaldehyde. A similar decreased activity can be seen in case of the *ortho* and *para* substituted benzaldehydes (Table 1). The presence of two Cl substitutions (inductive electron withdrawing groups) in 2,4-dichlorobenzaldehyde cause significant instability of the benzene ring which does not favour the reduction of the carbonyl group. Thus, both electronic and steric factors can explain the low activity of compound seen in entry 15, Table 1. Overall, the affinity of CpCR towards benzaldehyde decreases with increase in substitution on the benzene ring.

Table 2. Kinetics of CpCR.

Entry	Substrate	V_{max} (μmol min^{-1} mg^{-1})	K_m (mM)
1	Benzaldehyde	34 ± 0.9	0.29 ± 0.04
2	2, 4-dichlorobenzaldehyde	3.88 ± 0.56	7.14 ± 0.04

2.2. Oligomeric State of CpCR

The crystal structure of CpCR (PDB: 4OAQ) shows that the enzyme is a hetero dimer. SDS-PAGE indicates the presence of a 40 kDa sub unit in the buffer [19]. Gel filtration chromatography data indicates that CpCR is indeed a dimer of 80 kDa (Figure 1A) and remains so even at higher concentrations (Figure 1B). SRED from *Candida parapsilosis* ATCC 7330 exists in other oligomeric states unlike CpCR [29].

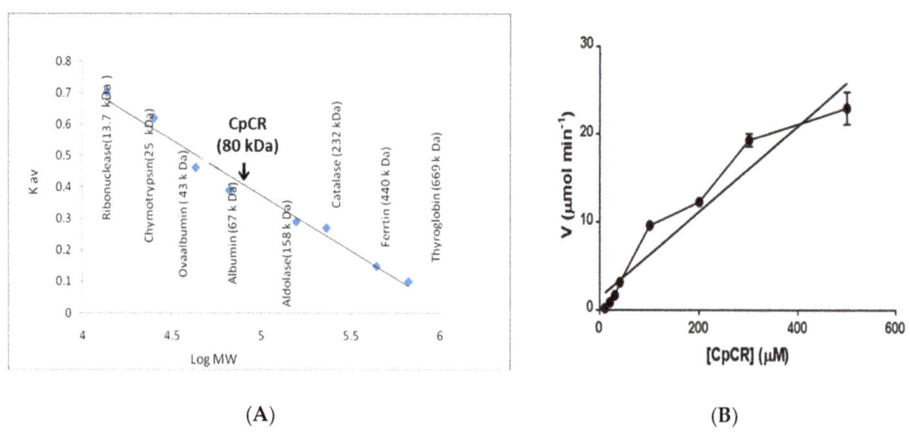

(A)   (B)

Figure 1. Oligomeric state of CpCR. (**A**) A plot of Log Mol. Wt. vs. K_{av} shows the calibration curve of the known standards including and the molecular weight of CpCR; (**B**) A plot of velocity of reaction vs. micro molar concentration of CpCR.

2.3. Effect of Chelating Agent EDTA and Divalent Metal Salts on Activity of CpCR

2.3.1. Effect of Time on Chelation

Addition of EDTA to the enzyme solution in a 1:3 enzyme: EDTA mole ratio, resulted in a drop of the specific activity of the enzyme immediately by 30% (from around 36 U/mg to 25 U/mg). Prolonged incubation of the mixture for up to 120 min showed that the specific activity remained around 25 U/mg for the entire duration. Furthermore, this experiment when repeated with a four-fold increased EDTA concentration (1:12—enzyme: EDTA mole ratio) over a lesser duration of 30 min showed the exact same trend (Figure S1). This indicates that EDTA binding to zinc in the enzyme is a fast process, and since the equilibrium is established, prolonged incubation is unnecessary.

2.3.2. Removal of EDTA

Different amounts of EDTA were incubated with a fixed concentration of enzyme in order to determine the effective concentration of EDTA necessary to remove zinc from the enzyme. However, it was found that despite the wide range of mole ratio tested, the specific activity of the protein samples containing EDTA remained the same (Figure S2). This was indicative of the fact that although EDTA binds to the enzyme quickly, it has to be removed in order to remove the zinc. This is consistent with what has been reported in literature [11,17,18].

The specific activities obtained after EDTA removal from samples by dialysis shows a decreasing trend in specific activity with increasing EDTA content (results not shown), indicating that the EDTA removal is necessary for zinc removal. A similar trend was observed in CPCR2 where the loss of activity of the enzyme is a function of loss of catalytic Zn [17]. However, the dialysis method was time consuming (12 h) and not feasible for this enzyme as it is not very stable. The specific activities obtained after EDTA removal from samples by ultrafiltration shows a similar decreasing trend using dialysis (Figure 2A). This method takes 3 h, and is likely to be associated with a gradual loss of specific activity which has to be taken into account. The protein recovery from this method is very high and, therefore, this method was optimized for use in future experiments.

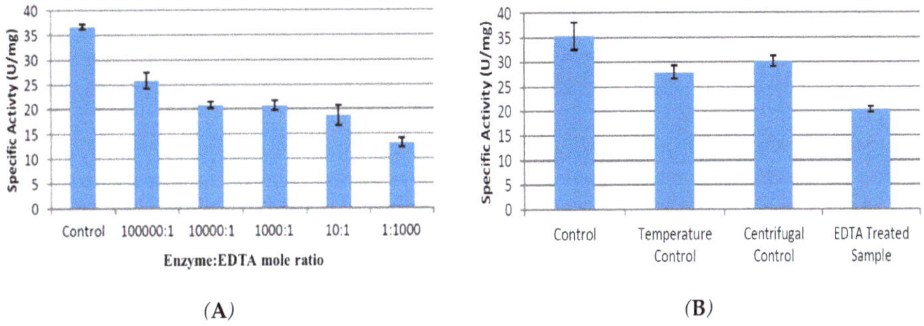

Figure 2. Activity of CpCR after chelation with EDTA (**A**) Effect of EDTA on CpCR activity at various concentrations; (**B**) Change in specific activity of CpCR before and after incubation at 4 °C.

Prolonged incubation during the EDTA removal on the enzyme sample was also studied. Two controls were designed to check spontaneous loss of activity—a temperature control that was placed at 4 °C for three hours and a centrifugal control that was placed in the filtration unit without EDTA treatment. The specific activity values were measured for all the controls and a sample treated with 10:1 enzyme:EDTA (Figure 2B). It was seen that there is a decrease of specific activity from 35 U/mg to 30 U/mg due to the three-hour incubation. However, the loss due to EDTA treatment was much more significant.

2.3.3. Inhibition of CpCR by Divalent Metal Salts

Attempts were made to restore specific activity of the enzyme samples treated with EDTA by addition of zinc chloride. Surprisingly, it was found that the addition of external zinc decreased the specific activity of not only the EDTA treated sample but also of the control sample without any EDTA (Figure 3A). This has been previously reported for carboxypeptidase-A and could be caused by bridging of the water molecule bound to the catalytic zinc to the external zinc thereby preventing substrate entry [30]. It was confirmed that this phenomenon is not just specific to zinc but to divalent ions of size similar to zinc as shown in Figure 3B. Reports on inhibition of ADHs by divalent metal ions suggest that the inhibition can be pH dependent and the mechanism mainly involves the replacement of native metal ion present in the catalytic site or by the coordination of added divalent metal ions with the sulphydryl groups of the enzyme. The inhibition can be reversed by addition of EDTA to remove the excess Zn ions [31–34]. Currently we are investigating the mechanism of inhibition of CpCR by such divalent metal salts.

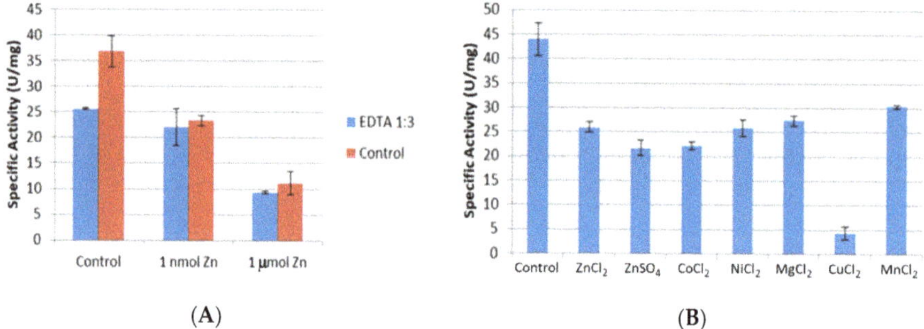

(A)   (B)

Figure 3. Effect of Zn^{2+} and other divalent metals on activity of CpCR. (**A**) Effect of Zn^{2+} ($ZnSO_4$) on CpCR activity at various concentrations; (**B**) Specific activity of CpCR influenced by various divalent metal ions.

2.4. Cofactor Pre-Treatment Prevents CpCR Activity Loss

Pre-treatment of enzyme samples with cofactors $NADPH/NADP^+$ prevented the loss of specific activity upon EDTA treatment in a concentration-dependent manner (Figure 4A). This is also reported with 1,10-phenanthroline when YADH is pre-incubated with the cofactor [18]. It was seen that pre-treatment with 1 mM $NADP^+$ gave 50% more specific activity as compared to the EDTA treated sample without cofactor treatment, while 2 mM NADPH retained the entire specific activity of the initial control. This may be due to retaining zinc and not allowing EDTA to access it possibly a result of structural changes due to cofactor binding.

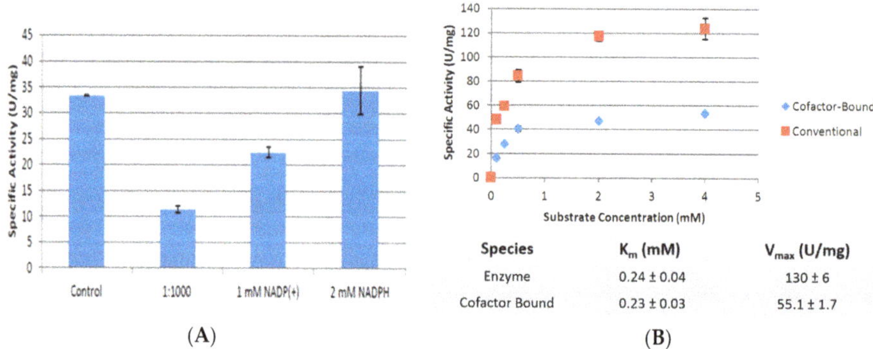

Species	K_m (mM)	V_{max} (U/mg)
Enzyme	0.24 ± 0.04	130 ± 6
Cofactor Bound	0.23 ± 0.03	55.1 ± 1.7

(A)   (B)

Figure 4. Effect of NADPH incubated CpCR on its activity. (**A**) Effect of incubation of different concentrations of NADPH on CpCR activity; (**B**) Kinetics of CpCR, NADPH bound vs. unbound.

The kinetics of the cofactor-bound enzyme were determined by the enzyme obtained from the modified purification protocol presented in this study with changes in the kinetic parameters duly noted. It was observed that the K_m value for both cofactor bound and unbound CpCR remained more or less constant at 0.23 mM, implying that substrate binding remained largely unchanged in the cofactor-bound enzyme. However, there was a two-fold decrease in V_{max} of cofactor-bound enzyme (Figure 4B). The structure of CpCR is significantly similar to LADH (PDB: 1HLD) with a *p*-value of 1.41×10^{-13} [35]. The decreased V_{max} value may also be attributed to the structural changes in the cofactor bound enzyme resulting in narrowing of the catalytic cleft and hindering the entry of the substrate [36]. Overall, the conversion of the *apo* enzyme to the *holo* form seems to affect the rate of the reaction significantly even though the enzyme affinity towards the substrate is retained.

3. Materials and Methods

3.1. Chemicals and Media

All the chemicals and media were purchased from SRL, Chennai, India and HiMedia, Mumbai, India. AKTA protein purification system and GST affinity column were purchased from GE Healthcare Life Sciences, Bangalore, India. The ultra-centrifugal filters were obtained from Merck, Mumbai, India.

3.2. Enzyme Expression and Purification

The overexpression and purification of CpCR were performed as per the reported methodology using GST affinity chromatography [28]. For all the experiments, except cofactor binding studies, the enzyme obtained from this protocol was used. Slight modifications to the protocol were done to obtain better yield and stability of the enzyme. They include: 1. Removal of DTT and $MgCl_2$ from all the buffers and replacing the Tris HCl with HEPES buffer. 2. Cell disruption was done using a 150 W ultra-sonicator. 3. The cleared lysate was loaded onto the GST column at a flow rate of 1 ml min^{-1}. 4. The use of a superdex column was skipped. The yield and fold purifications were obtained by checking the specific activity of CpCR against benzaldehyde.

The composition of the buffers used are as follows: Equilibration and wash buffer: 50 mM HEPES pH 7.5, 500 mM NaCl, 2.5% glycerol; Elution buffer: 50 mM HEPES pH 7.5, 20 mM Glutathione; Desalting buffer: 20 mM HEPES pH 7.5, 50 mM NaCl.

3.3. Specific Activity, Substrate Scope and Kinetic Studies of CpCR

The protocol used for determining the specific activity of the enzyme against different substrates was the same as reported previously [3]. The activity of CpCR against benzaldehyde substituents and aliphatic aldehydes was checked. Substrate concentrations varying from 0–4 mM were used to determine the specific activity of CpCR and the Lineweaver–Burk plot gave the K_m and V_{max} of CpCR against the specific substrates.

3.4. Oligomeric State of CpCR

Gel filtration chromatography was done to find out the oligomeric state of CpCR [37]. Mixture of standard proteins containing ribonuclease (13.7 kDa), chymotrypsin (25 kDa), ovalbumin (43 kDa), albumin (67 kDa), aldolase (158 kDa), catalase (232 kDa), ferritin (440 kDa) and thyroglobulin (669 kDa) were used to calibrate the Sephadex 200 HR column. A plot of Log Mol. Wt. and K_{av} was made to calculate the molecular weight of CpCR.

A plot of velocity (μmoles min^{-1}) vs. concentration of CpCR (μM) was made to find the presence of higher oligomeric state of CpCR at its higher concentrations of up to 500 μM.

3.5. Treatment of CpCR with EDTA, Divalent Metal Salts

Benzaldehyde was used as the substrate to check the activity of CpCR against the effects of EDTA and other metal salts. The mixture containing the ratio of 1:3 and 1:12 (number of moles of CpCR to the number of moles of EDTA) was checked for the activity instantaneously and compared to the untreated CpCR. Another experiment with the same mixtures incubated for up to two hours at 4 °C with the activity checked every 15 min from the start of incubation was also done. Additionally, different ratios of the number of moles of CpCR to EDTA were tried to see the instantaneous effect on the activity of CpCR.

Ultrafiltration was done to remove the low Mol. Wt. EDTA from the above mixtures to later determine the change in the activity of CpCR. The chelator treated protein (100–500 µg) in a volume < 500 µL was placed in 0.5 mL filter (Amicon, 10 kDa mol. wt. cutoff) and the volume was made up to 500 µL using a desalting buffer. The samples were then concentrated at 14,000 g at 4° C for 30 min, following which the volume was again made up to 500 µL. This process was repeated for a total of six times over three hours. At the end of the process, the enzyme was recovered and its concentration was determined by Bradford's method [38] followed by its activity assay.

The addition of micro- and nanomolar concentrations of $ZnSO_4$ to the EDTA treated CpCR was also studied. The metal salt in question was directly added to the assay solution from a 1 M stock to obtain the desired concentration after addition of substrate and enzyme to determine the activity spectrophotometrically. Effects of divalent metal salts like $CoCl_2$, $NiCl_2$, $MgCl_2$, $CuCl_2$, $MnCl_2$ and $ZnCl_2$ at nanomolar concentrations on the CpCR activity were noted.

3.6. CpCR—Cofactor Binding Studies

For this study, the modified protocol for purification was used. NADPH was added to 100 µg of the protein from 10 mM and 20 mM stock of cofactor to a final concentration of 1 mM and 2 mM respectively, and the solution was incubated at 4 °C for one hour. Following this, the sample was treated with EDTA maintaining a 1:1000 enzyme: EDTA mole ratio. The EDTA was removed by ultra-centrifugation, and the specific activity of the sample was determined. Kinetic studies of CpCR against benzaldehyde were compared with the CpCR pre-incubated with 2 mM NADPH.

4. Conclusions

CpCR substrates include substituted benzaldehydes and aliphatic aldehydes. Substituted benzaldehydes showed lower activity as compared to benzaldehyde. The oligomeric state of the enzyme was confirmed to be dimeric at all concentrations, in agreement with the crystal structure. In chelation studies with EDTA, a decrease in enzyme activity with an increase in EDTA concentration is seen after removal of the EDTA from the solution by ultrafiltration. All the divalent metal ions inhibit the activity of CpCR even at nanomolar concentrations. The protocol for CpCR purification was modified to obtain 10 times more yield and > 3-fold purification and used in studying the cofactor binding studies. The pre-incubation of CpCR with cofactor makes the enzyme resist the Zn removal by EDTA chelation and retains activity. The *apo* and *holo* forms of CpCR do not differ in their affinity towards benzaldehyde but differ in their reaction rates.

Supplementary Materials: The following are available online at http://www.mdpi.com/2073-4344/9/9/702/s1; Figure S1. Time study of chelation; Figure S2. Concentration based chelation without removing the chelator.

Author Contributions: Conceptualization, A.C., V.K.K. and D.C.; Experiments carried out by, V.K.K. and D.C.; Formal analysis, A.C., V.K.K. and D.C.; Writing—original draft preparation, V.K.K.; Writing—review and editing, A.C.; Supervision, A.C.

Funding: This research received no external funding.

Acknowledgments: One of the authors, Vinay Kumar Karanam, expresses his gratitude to IIT Madras, India, for the fellowship.

Conflicts of Interest: The authors declare no conflict of interest.

References

1. Persson, B.; Hedlund, J.; Jörnvall, H. Medium- and short-chain dehydrogenase/reductase gene and protein families: The MDR superfamily. *Cell. Mol. Life Sci.* **2008**, *65*, 3879–3894. [CrossRef] [PubMed]

2. Jörnvall, H.; Hedlund, J.; Bergman, T.; Oppermann, U.; Persson, B. Biochemical and Biophysical Research Communications Superfamilies SDR and MDR: From early ancestry to present forms. Emergence of three lines, a Zn-metalloenzyme, and distinct variabilities. *Biochem. Biophys. Res. Commun.* **2010**, *396*, 125–130. [CrossRef] [PubMed]

3. Aggarwal, N.; Ananthathamula, R.; Karanam, V.K.; Doble, M.; Chadha, A. Understanding substrate specificity and enantioselectivity of carbonyl reductase from *Candida parapsilosis* ATCC 7330 (CpCR): Experimental and modeling studies. *Mol. Catal.* **2018**, *460*, 40–45. [CrossRef]

4. Chadha, A.; Venkataraman, S.; Preetha, R.; Padhi, S.K. *Candida parapsilosis*: A versatile biocatalyst for organic oxidation-reduction reactions. *Bioorg. Chem.* **2016**, *68*, 187–213. [CrossRef] [PubMed]

5. Kvassman, J.; Pettersson, G. Kinetic Transients in the Reduction of Aldehydes Catalysed by Liver Alcohol Dehydrogenase. *Eur. J. Biochem.* **1976**, *69*, 279–287. [CrossRef] [PubMed]

6. Deetz, J.S.; Luehr, C.A.; Vallee, B.L. Human Liver Alcohol Dehydrogenase Isozymes: Reduction of Aldehydes and Ketones. *Biochemistry* **1984**, *23*, 6822–6828. [CrossRef]

7. Pal, S.; Park, D.H.; Plapp, B.V. Activity of yeast alcohol dehydrogenases on benzyl alcohols and benzaldehydes. Characterization of ADH1 from *Saccharomyces carlsbergensis* and transition state analysis. *Chem. Biol. Interact.* **2009**, *178*, 16–23. [CrossRef]

8. Jelokova, J.; Karlsson, C.; Estonius, M.; Jornvall, H.; Hoog, J.O. Features of structural zinc in mammalian alcohol dehydrogenase. Site-directed mutagenesis of the zinc ligands. *Eur. J. Biochem.* **1994**, *225*, 1015–1019. [CrossRef]

9. Magonet, E.; Hayen, P.; Delforge, D.; Delaive, E.; Remacle, J. Importance of the structural zinc atom for the stability of yeast alcohol dehydrogenase. *Biochem. J.* **1992**, *287 Pt 2*, 361–365. [CrossRef]

10. Auld, D.S.; Bergman, T. Medium- and short-chain dehydrogenase/reductase gene and protein families: The role of zinc for alcohol dehydrogenase structure and function. *Cell. Mol. Life Sci.* **2008**, *65*, 3961–3970. [CrossRef]

11. Kägi, J.H.R.; Vallee, B.L. The Role of Zinc in Alcohol Dehydrogenase. *J. Biol. Chem.* **1960**, *235*, 3188–3192.

12. Baker, P.J.; Britton, K.L.; Fisher, M.; Esclapez, J.; Pire, C.; Bonete, M.J.; Ferrer, J.; Rice, D.W. Active site dynamics in the zinc-dependent medium chain alcohol dehydrogenase superfamily. *Proc. Natl. Acad. Sci. USA* **2009**, *106*, 779–784. [CrossRef]

13. Drum, D.E.; Harrison, J.H.; Li, T.K.; Bethune, J.L.; Vallee, B.L. Structural and functional zinc in horse liver alcohol dehydrogenase. *Proc. Natl. Acad. Sci. USA* **1967**, *57*, 1434–1440. [CrossRef]

14. Vallee, B.L.; Auld, D.S. Active-site zinc ligands and activated H2O of zinc enzymes. *Proc. Natl. Acad. Sci. USA* **1990**, *87*, 220–224. [CrossRef]

15. Ryde, U. The coordination chemistry of the structural zinc ion in alcohol dehydrogenase studied by ab initio quantum chemical calculations. *Eur. Biophys. J.* **1996**, *24*, 213–221. [CrossRef]

16. Wang, J.; Sakakibara, M.; Matsuda, M.; Itoh, N. Site-directed Mutagenesis of Two Zinc- binding Centers of the NADH-dependent Phenylacetaldehyde Reductase from Styrene- assimilating *Corynebacterium* sp. Strain ST-10. *Biosci. Biotechnol. Biochem.* **2014**, *63*, 2216–2218. [CrossRef]

17. Man, H.; Loderer, C.; Ansorge-Schumacher, M.B.; Grogan, G. Structure of NADH-dependent carbonyl reductase (CPCR2) from *Candida parapsilosis* provides insight into mutations that improve catalytic properties. *ChemCatChem* **2014**, *6*, 1103–1111. [CrossRef]

18. Dickinson, F.M.; Berrieman, S. The reactions of 1,10-phenanthroline with yeast alcohol dehydrogenase. *Biochem. J.* **1977**, *167*, 237–244. [CrossRef]

19. Dołega, A. Alcohol dehydrogenase and its simple inorganic models. *Coord. Chem. Rev.* **2010**, *254*, 916–937. [CrossRef]

20. Chen, F.; Wang, P.; An, Y.; Huang, J.; Xu, Y. Structural insight into the conformational change of alcohol dehydrogenase from *Arabidopsis thaliana* L. during coenzyme binding. *Biochimie* **2014**, *108*, 33–39. [CrossRef]

21. Colonna-Cesari, F.; Perahia, D.; Karplus, M.; Eklund, H.; Bräden, C.I.; Tapia, O. Interdomain motion in liver alcohol dehydrogenase. Structural and energetic analysis of the hinge bending mode. *J. Biol. Chem.* **1986**, *261*, 15273–15280.

22. Marolt, M.; Lüdeke, S. Studying NAD(P)H cofactor-binding to alcohol dehydrogenases through global analysis of circular dichroism spectra. *Phys. Chem. Chem. Phys.* **2019**, *21*, 1671–1681. [CrossRef]

23. Satheesan Babu, C.; Lim, C. Efficient Binding of Flexible and Redox-Active Coenzymes by Oxidoreductases. *ACS Catal.* **2016**, *6*, 3469–3472. [CrossRef]

24. Thompson, M.P.; Turner, N.J. Two-Enzyme Hydrogen-Borrowing Amination of Alcohols Enabled by a Cofactor-Switched Alcohol Dehydrogenase. *ChemCatChem* **2017**, *9*, 3833–3836. [CrossRef]

25. Cahn, J.K.B.; Werlang, C.A.; Baumschlager, A.; Brinkmann-Chen, S.; Mayo, S.L.; Arnold, F.H. A General Tool for Engineering the NAD/NADP Cofactor Preference of Oxidoreductases. *ACS Synth. Biol.* **2017**, *6*, 326–333. [CrossRef]

26. Chen, H.; Zhu, Z.; Huang, R.; Zhang, Y.H.P. Coenzyme Engineering of a Hyperthermophilic 6-Phosphogluconate Dehydrogenase from NADP+ to NAD+ with Its Application to Biobatteries. *Sci. Rep.* **2016**, *6*, 1–8. [CrossRef]

27. You, C.; Huang, R.; Wei, X.; Zhu, Z.; Zhang, Y.H.P. Protein engineering of oxidoreductases utilizing nicotinamide-based coenzymes, with applications in synthetic biology. *Synth. Syst. Biotechnol.* **2017**, *2*, 208–218. [CrossRef]

28. Aggarwal, N.; Mandal, P.K.; Gautham, N.; Chadha, A. Expression, purification, crystallization and preliminary X-ray diffraction analysis of carbonyl reductase from *Candida parapsilosis* ATCC 7330. *Acta Crystallogr. Sect. F Struct. Biol. Cryst. Commun.* **2013**, *69*, 313–315. [CrossRef]

29. Sudhakara, S.; Chadha, A. A carbonyl reductase from *Candida parapsilosis* ATCC 7330: Substrate selectivity and enantiospecificity. *Org. Biomol. Chem.* **2017**, *15*, 4165–4171. [CrossRef]

30. Larsen, K.S.; Auld, D.S. Carboxypeptidase A: Mechanism of Zinc Inhibition. *Biochemistry* **1989**, *28*, 9620–9625. [CrossRef]

31. Maret, W.; Yetman, C.A.; Jiang, L.J. Enzyme regulation by reversible zinc inhibition: Glycerol phosphate dehydrogenase as an example. *Chem. Biol. Interact.* **2001**, *130*, 891–901. [CrossRef]

32. Ying, X.; Wang, Y.; Badiei, H.R.; Karanassios, V.; Ma, K. Purification and characterization of an iron-containing alcohol dehydrogenase in extremely thermophilic bacterium *Thermotoga hypogea*. *Arch. Microbiol.* **2007**, *187*, 499–510. [CrossRef]

33. Stiborová, M.; Leblová, S. Effect of Metals on Rape Alcohol Dehydrogenase. *Biochem. Physiol. Pflanz.* **2017**, *174*, 39–43. [CrossRef]

34. Jin, L.; Szeto, K.Y.; Zhang, L.; Du, W.; Sun, H. Inhibition of alcohol dehydrogenase by bismuth. *J. Inorg. Biochem.* **2004**, *98*, 1331–1337. [CrossRef]

35. Ye, Y.; Godzik, A. FATCAT: A web server for flexible structure comparison and structure similarity searching. *Nucleic Acids Res.* **2004**, *32*, 582–585. [CrossRef]

36. Brändén, C.; Eklund, H. Coenzyme-induced Conformational Changes and Substrate Binding in Liver Alcohol Dehydrogenase. In *CIBA Foundation Symposium 60—Molecular Interactions and Activity in Proteins*; Porter, R., Fitzsimons, D.W., Eds.; Ciba Foundation: London, UK, 2009; pp. 63–80.

37. Aggarwal, N. Cloning, Purification, Biochemical Characterisation and Crystallisation of a Carbonyl Reductase from *Candida parapsilosis* ATCC 7330: Towards Understanding Its Enantioselectivity at the Molecular Level. Ph.D. Thesis, Indian Institute of Technology Madras, Chennai, India, 2013.

38. Bradford, M.M. A rapid and sensitive method for the quantitation of microgram quantities of protein utilizing the principle of protein-dye binding. *Anal. Biochem.* **1976**, *72*, 248–254. [CrossRef]

Article

Accelerated H$_2$ Evolution during Microbial Electrosynthesis with *Sporomusa ovata*

Pier-Luc Tremblay [1,2,†], Neda Faraghiparapari [3,†] and Tian Zhang [1,2,3,*]

[1] State Key Laboratory of Silicate Materials for Architectures, Wuhan University of Technology, Wuhan 430070, China; pierluct@whut.edu.cn
[2] School of Chemistry, Chemical Engineering and Life Science, Wuhan University of Technology, Wuhan 430070, China
[3] The Novo Nordisk Foundation Center for Biosustainability, Technical University of Denmark Kemitorvet-220, 2800 Lyngby, Denmark; nfaraghi@gmail.com
*   Correspondence: tzhang@whut.edu.cn; Tel.: +86-181-8643-6590
†   These two authors contributed equally to this work.

Received: 14 January 2019; Accepted: 1 February 2019; Published: 8 February 2019

Abstract: Microbial electrosynthesis (MES) is a process where bacteria acquire electrons from a cathode to convert CO$_2$ into multicarbon compounds or methane. In MES with *Sporomusa ovata* as the microbial catalyst, cathode potential has often been used as a benchmark to determine whether electron uptake is hydrogen-dependent. In this study, H$_2$ was detected by a microsensor in proximity to the cathode. With a sterile fresh medium, H$_2$ was produced at a potential of −700 mV versus Ag/AgCl, whereas H$_2$ was detected at −500 mV versus Ag/AgCl with cell-free spent medium from a *S. ovata* culture. Furthermore, H$_2$ evolution rates were increased with potentials lower than −500 mV in the presence of cell-free spent medium in the cathode chamber. Nickel and cobalt were detected at the cathode surface after exposure to the spent medium, suggesting a possible participation of these catalytic metals in the observed faster hydrogen evolution. The results presented here show that *S. ovata*-induced alterations of the cathodic electrolytes of a MES reactor reduced the electrical energy required for hydrogen evolution. These observations also indicated that, even at higher cathode potentials, at least a part of the electrons coming from the electrode are transferred to *S. ovata* via H$_2$ during MES.

Keywords: industrial biotechnology; electrochemistry; biohydrogen; biocatalysis; process development; bacteria

---

1. Introduction

Reductive bioelectrochemical processes rely on the transfer of electrons from a cathode to a microbial catalyst for the reduction of a substrate with protons coming from an anodic reaction [1,2]. The substrate can be inorganic carbon molecules like CO$_2$ that will be reduced to multicarbon compounds or CH$_4$ via microbial electrosynthesis (MES) [3–10]. Organic carbon compounds can also be converted into commodity chemicals via electrofermentation [11,12] or electrorespiration [13].

In reductive bioelectrochemical systems (BES), the electrons are thought to be transferred directly via physical contact between the microbes and the cathode or indirectly via an electron shuttle such as H$_2$ [14–16]. Experimental evidences suggest that H$_2$ evolution from a graphite electrode often used in reductive BES starts happening only at potentials below −800 mV vs. Ag/AgCl in batch experiments [17]. Thus, it has been proposed that when the cathode potential is set higher than −800 mV, electrons are transferred via a H$_2$-independent mechanism that could possibly involve the direct acquisition of electrons by components of the bacterium cell wall [3,18].

A recent study indicated that in the presence of cell-free spent medium from the electroactive acetogen *Sporomusa sphaeroides*, a significant quantity of H_2 is produced in a BES with a cathode potential set at -710 mV vs. Ag/AgCl [19]. Furthermore, the same study showed that in the presence of cell-free spent medium from the electroactive methanogen *Methanococcus maripaludis*, H_2, as well as formate, are formed in a BES at higher cathode potential compared to sterile fresh medium control and in sufficient quantities to account for all the methane produced from CO_2. The authors suggested a novel electron transfer mechanism in which hydrogenases and formate dehydrogenase released in the medium from microbial cells would interact with a cathode set at a potential above -800 mV vs. Ag/AgCl to catalyze the formation of soluble electron shuttles. Alternatively, other groups have proposed that copper, nickel, iron, or vanadium deposited at the surface of a cathode via microbial activity could be responsible for the increase of bioelectrochemical hydrogen production observed at different potentials after exposure of the cathode to microbial catalysts [20–22].

Cathode materials [23], reactor designs [24–27], and operating modes [28,29] are all parameters that can positively affect H_2 evolution. The chemistry of the solution filling the cathodic reactor also has an impact on the relation between the cathode potential and H_2 evolution through changes in H_2 initial concentration [22,30], changes in buffer composition [31–33], and through the presence of weak acids [34,35]. During MES, the microbial catalyst will alter the chemical environment surrounding the cathode by releasing metabolic wastes, products, or diverse cell components or debris in the cathodic solution [19,21,36,37]. To study the possible correlation between microbial alterations and H_2 evolution in a MES system, a hydrogen microsensor was placed in close proximity to the surface of a cathode set at different potentials to measure H_2 evolution in situ in the presence of sterile medium, bacterial culture or cell-free filtrate. *Sporomusa ovata*, one of the best pure culture MES catalysts for the production of acetate from CO_2 [38,39] is used here as a model because of its capacity to perform MES over a large range of cathode potential [3,40–46]. In order to further understand H_2 evolution and electron uptake during MES, other variables were investigated, including the presence of metals at the cathode surface as well as the presence of hydrogenases and other enzymes in the cell-free spent medium.

2. Results and Discussion

2.1. H₂ Evolution in an Abiotic MES Reactor

A hydrogen microsensor with a sensitivity of ≥ 0.1 µM was inserted into an abiotic MES reactor to monitor H_2 evolution with a cathode set at potential ranging from -900 to -400 mV vs. Ag/AgCl (Figure S1). H_2 concentration was measured in close proximity to the cathode surface where microbial catalysts in operating MES reactor [3,38,41] are likely to oxidize large fraction of H_2, if any is produced, before it can diffuse away in the medium and in the reactor gas phase. The initial evaluation of H_2 evolution was conducted in an abiotic MES reactor. Both cathodic and anodic chambers were filled with sterile 311 medium at pH 6.8, which is the growth medium as well as the electrolyte solution normally used in *S. ovata*-driven MES reactors [3]. Under these experimental conditions, the highest cathode potential at which H_2 evolution was observed was -700 mV vs. Ag/AgCl (Table 1). As expected, more current was drawn at lower potentials versus higher potentials because H_2 evolved faster at lower potentials (Figure 1, Table 1, Figure S2).

Table 1. H_2 evolution, current density and electrons recovery in MES reactors with sterile fresh 311 medium [a].

Potential vs. Ag/AgCl (mV)	H₂ Evolution Rates (µM min⁻¹)	Current Density (mA m⁻²)	Electrons Recovery in H₂ (%)
−900	0.088 ± 0.012	-19.8 ± 2.1	87 ± 1.8
−850	0.073 ± 0.010	-16.3 ± 1.7	85 ± 3.7
−800	0.059 ± 0.008	-14.9 ± 1.8	80 ± 2.1
−750	0.052 ± 0.011	-13.7 ± 1.2	79 ± 9.0
−700	0.048 ± 0.007	-12.8 ± 1.0	78 ± 7.0
−600	n.d. [b]	n/a [c]	n/a

[a] Each result is the mean and standard deviation of three replicates. [b] Not detected. [c] Not applicable.

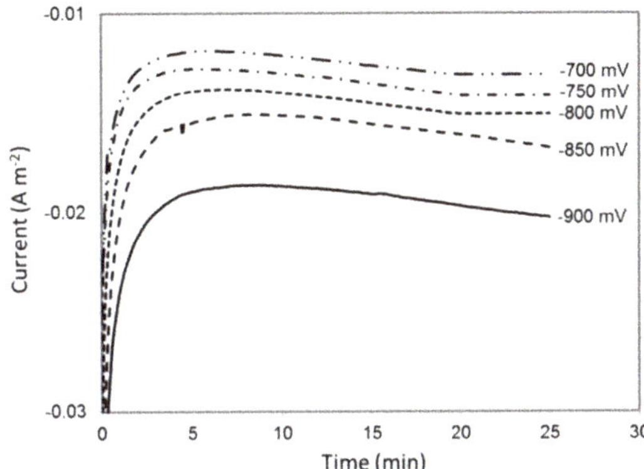

Figure 1. Current draw in a MES reactor filled with sterile fresh 311 medium at different cathode potentials. Results shown are from a representative example of three replicate.

2.2. H_2 Evolution in the Presence of a S. ovata Cell Suspension

S. ovata is an efficient acetogenic microbial catalyst for the production of acetate from CO_2 by MES capable of acquiring electrons from the cathode at potentials as high as −600 mV vs. Ag/AgCl [1,3,41]. To determine the impact of *S. ovata* on H_2 accumulation in a MES system, a cell suspension was added to a cathode chamber equipped with a H_2 microsensor. At the tested potentials higher than −900 mV, no H_2 was detected (Table 2). At −900 mV, 0.11 ± 0.02 µM min⁻¹ (n = 3) of H_2 accumulated, and the electron recovery from current to H_2 was only $8.5 \pm 1.5\%$ (Table 2, Figure 2, Figure S3), indicating that *S. ovata* cells probably quickly consumed most of the H_2 generated at the cathode. In the meantime, ca. 6.6 ± 0.2 µM of acetate was produced by *S. ovata* cell suspension from CO_2 in the cathode chamber of the MES system over a period of 25 minutes, and the coulombic efficiency was $75 \pm 3\%$.

Table 2. H_2 evolution, current density and electrons recovery in MES reactors with *S. ovata* cell suspension [a].

Potential vs. Ag/AgCl (mV)	H₂ Evolution Rates (µM min⁻¹)	Current Density (mA m⁻²)	Electrons Recovery in H₂ (%)
−900	0.11 ± 0.02	-317 ± 32	8.5 ± 1.5
−850	n.d. [b]	n/a [c]	n/a
−800	n.d.	n/a	n/a
−750	n.d.	n/a	n/a
−700	n.d.	n/a	n/a
−600	n.d.	n/a	n/a

[a] Each result is the mean and standard deviation of three replicates. [b] Not detected. [c] Not applicable.

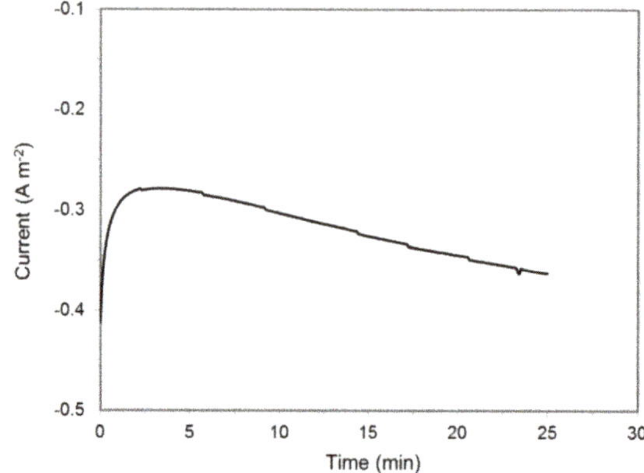

Figure 2. Current draw in a MES reactor containing a *S. ovata* cell suspension at a cathode potential of −900 mV vs. Ag/AgCl. Results shown are from a representative example of three replicate.

2.3. H_2 Evolution Shifting in the Presence of a Cell-Free Filtrate from S. ovata Culture

To recreate the chemical environment of an operating MES reactor in the absence of the H_2-oxidizing microbial catalyst, a cell-free filtrate from a *S. ovata* culture grown beforehand for four weeks under electrosynthetic condition was employed as the cathodic solution. Compared to a *S. ovata* cell suspension, H_2 was present in significant quantities at −900 mV vs. Ag/AgCl with simultaneous current draw indicating that the cell-free filtrate was unable to oxidize detectable amount of H_2 (Table 3, Figure 3). Furthermore, higher H_2 evolution rates in the MES reactor were detected with the cell-free filtrate than with the sterile fresh medium at cathode potential ranging from −900 mV to −700 mV vs. Ag/AgCl (Figure 4, Figure S4). In addition, the detectable H_2 evolution in the presence of cell-free filtrate was shifted by +200 mV compared to sterile medium control. H_2 started to accumulate at −500 mV with *S. ovata* cell-free filtrate, whereas H_2 evolution was detected at −700 mV with sterile fresh 311 medium. These results suggested that secreted metabolites, secreted cell components or chemical characteristics of the *S. ovata* cell-free filtrate enabled H_2 evolution at higher cathode potentials and accelerated it. Furthermore, it seems to indicate that electron transfer in MES driven by *S. ovata* at high cathode potentials involved H_2 as an electron shuttle. However, further characterization is required to determine whether all the electrons required for the acetate production by MES at high cathode potentials are transferred via H_2, or if a significant fraction of the electrons comes from an alternative H_2-independent route such as direct electron transfer.

Table 3. H_2 evolution, current density and electrons recovery in MES reactors with cell-free spent medium [a].

Potential vs. Ag/AgCl (mV)	H_2 eVolution Rates (µM min^{-1})	Current Density (mA m^{-2})	Electrons Recovery in H_2 (%)
−900	1.91 ± 0.25	−398 ± 90	87 ± 8.4
−850	1.71 ± 0.14	−320 ± 86	89 ± 8.0
−800	1.34 ± 0.10	−289 ± 38	85 ± 7.9
−750	1.17 ± 0.17	−276 ± 51	80 ± 3.2
−700	1.03 ± 0.12	−254 ± 60	75 ± 3.0
−600	0.84 ± 0.10	−225 ± 47	74 ± 5.9
−500	0.73 ± 0.08	−179 ± 42	77 ± 9.0
−400	n.d. [b]	n/a [c]	n/a

[a] Each result is the mean and standard deviation of three replicates. [b] Not detected. [c] Not applicable.

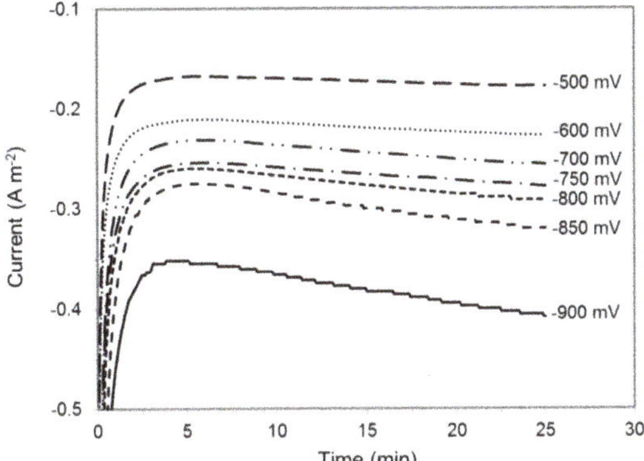

Figure 3. Current draw in a MES reactor containing cell-free spent medium from an electrosynthetic *S. ovata* culture. Results shown are from a representative example of three replicate.

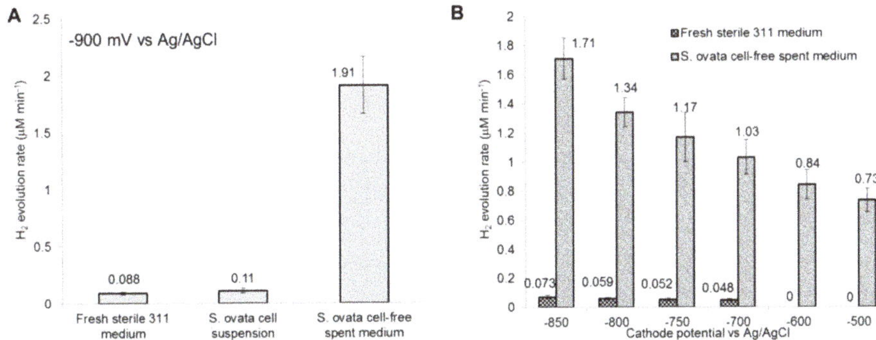

Figure 4. H_2 evolution rate with fresh sterile medium, *S. ovata* cell suspension and cell-free spent medium. (**A**) Cathode potential at -900 mV vs. Ag/AgCl and (**B**) cathode potentials ranging from -850 to -500 mV. Above -500 mV vs. Ag/AgCl, no H_2 evolution was detected under all tested conditions. No H_2 evolution was detected with *S. ovata* cell suspension above -900 mV vs. Ag/AgCl. Each result is the mean and standard deviation of three replicates.

A possible explanation for the faster H_2 evolution observed here would be the presence of acetic acid in the cell-free spent medium. Weak acids including acetic acid have been shown to have a catalytic effect on H_2 evolution in BES at acidic pH as well as in abiotic electrochemical systems [34,35]. In the cell-free filtrate samples of *S. ovata* tested here, the concentration of acetate/acetic acid varied from 10.4 to 12.4 mM (Figure S5). This was generated by the filtered cells beforehand during the MES process. However, the pH of the cell-free filtrate maintained with a carbonate buffer was measured at ca. 6.8, which suggests that the acetic acid/acetate ratio is unlikely to be high enough to have a significant impact on the electrochemical generation of H_2.

2.4. Metals at the Surface of the Cathode after Exposure to Cell-Free Spent Medium

Energy-dispersive X-ray spectroscopy (EDS) was employed to examine the presence of elements on the surface of cathode electrodes exposed to electrosynthetic *S. ovata* cell-free spent medium or sterile 311 medium. As expected, C, O, Na, Mg, P, Cl, K and Ca were observed on spectra for each condition (Figure 5). Interestingly, two metal elements, i.e., cobalt and nickel, that may be involved in the acceleration of H_2 evolution, were detected with the electrode samples exposed to cell-free spent medium but not with samples exposed to sterile 311 medium. When subtracting the background, the EDS signal corresponding to cobalt had an X-ray count of ca. 50 and 120 in cell-free spent medium replicate A and B, respectively. For nickel, the X-ray count was ca. 60 for replicate A and 140 for replicate B. No signal corresponding to nickel or cobalt was detected with sterile 311 medium (Figure 5C).

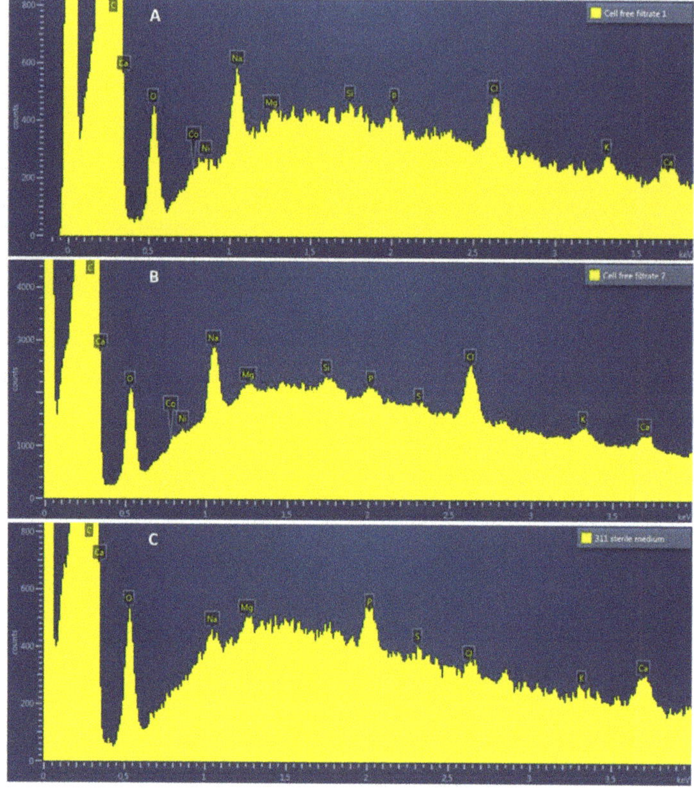

Figure 5. EDS spectra (from 0.0 to 4 KeV) of the cathode electrode surface after exposure to cell-free spent medium from electrosynthetic *S. ovata* (**A,B**) or to sterile fresh 311 medium (**C**).

Multiple proteins and enzymes found in acetogens, such as NiFe hydrogenases, acetyl-CoA synthase, CO dehydrogenase, corrinoid iron-sulfur protein and cobalamin-dependent methyltransferases, have metal centers containing nickel or cobalt [47–49]. Evidence suggested that hydrogenases released in the medium by S. sphaeroides cells are responsible for improved H_2 evolution from a cathode [19]. In this study, we were unsuccessful at detecting hydrogenase activity in the cell-free spent medium with a methyl-viologen based assay. Additionally, SDS-PAGE and mass spectrometry were applied to examine *S. ovata* cell-free spent medium and no intact hydrogenases were detected (Figure S6). Further research is warranted to establish if functional hydrogenases or other redox enzymes are released by *S. ovata* during MES, and if these enzymes participates actively to electron transfer from the cathode.

Three main visible bands were excised from the SDS-PAGE corresponding to an aldehyde oxidoreductase (Sov_1c12660), a glutamate synthase (gltB2) and an ll-diaminopimelate aminotransferase (dapL2) (Figure S6). All three are predicted to be cytoplasmic proteins [50], which suggest that microbial cell content was spilled in the MES reactor medium due to lysis. This further increased the chemical complexity of the material surrounding the cathode. The presence of cytoplasmic enzymes in the cell-free spent medium suggests that the detected nickel and cobalt at the surface of the cathode could come from three sources: intact enzymes, metal centers attached to damaged enzymes and free metal centers detached from apoenzymes. Furthermore, metal centers associated for instance with Sov_1c12660 (Fe, Mo), as well as with other proteins possibly found in the cell-free spent medium of *S. ovata*, could also interact transiently with the cathode and facilitate hydrogen evolution.

3. Materials and Methods

3.1. Bacterium and Growth Conditions

S. ovata DSM-2662 [51] was obtained from the Deutsche Sammlung Mikroorganismen und Zellkulturen (DSMZ, Braunschweig, Germany). *S. ovata* strains were routinely maintained in the DSMZ-recommended 311 medium at 30 °C and at pH 6.8 with a $H_2:CO_2$ (80:20) atmosphere (1.7 atm). Casitone, sodium sulfide, yeast extract, and resazurin were omitted from the 311 medium. For MES experiments, cysteine was also omitted from the 311 medium.

3.2. Preparation of Cell Suspension

Triplicate of 300 mL of anoxic cultures of *S. ovata* were harvested by centrifugation and washed two times with 311 sterile medium before being resuspended in a final volume of 3 mL. $H_2:CO_2$-grown *S. ovata* cells were harvested when the optical density was ca. 0.3 (545 nm). Cell suspensions were then used to inoculate the cathodic chamber of MES reactors containing 250 mL of sterile 311 medium.

3.3. Cell-Free Spent Medium of S. ovata

S. ovata cultures at an OD545 of 0.2 catalyzing the conversion of CO_2 to acetate for 4 weeks in a MES reactor with a cathode potential of -900 mV vs. Ag/AgCl/3M KCl were filtered two times with 0.45 μm pore size filters. 250 mL of cell-free spent medium was then injected in the cathodic chamber of a MES reactor.

3.4. MES Reactor and H$_2$ Evolution

Three-electrode H-cell glass bioreactor (Adams and Chittenden Scientific Glass, Berkeley, CA, USA) systems were used for H_2 evolution experiments during MES. The anode chamber was filled with 250 mL of sterile 311 medium, whereas the cathode chamber contained S. ovata cell suspension, *S. ovata* cell-free spent medium or sterile 311 medium. The anode and cathode chambers were separated by a Nafion 115 membrane (DuPont Inc., Wilmington, DE, USA). Graphite plates (type HLM, SGL carbon, Wiesbaden, Germany) with a normalized surface area of 35.5 cm^2 were used as both anode and cathode. The anode and cathode chamber was stirred at 300 rpm and bubbled with $N_2:CO_2$ (80:20) gas mixture at a flow rate of 18.5 $\pm$ 1.0 mL/min (ADM 2000 Flowmeter, Agilent, Santa Clara, CA, USA). For the data presented here, the stirring and bubbling in cathode chamber was paused for 25 min during H_2 measurement to avoid interference of accuracy measurement. Fixed potentials were applied to the cathode from -900 to -400 mV vs. Ag/AgCl using a multi-potentiostat (CHI 1000C, CH Instrument, Inc., Austin, TX, USA).

H_2 evolution was measured in close proximity to the surface of the graphite cathode of a MES reactor maintained at 25 °C with a hydrogen microsensor (H2-500, Unisense, Aarhus, Denmark). The microsensor with a tip surface area of 0.2 mm^2 was installed in the MES reactor and the distance between the tip of the H_2 microsensor and the cathode surface was ca. 2 mm (Figure S1). Before each experiment, the microsensor was calibrated with a gas mixture containing 7% H_2. The sensor data was logged continuously every second using a data logger (Microsensor Multimeter, Unisense, Aarhus, Denmark). H_2 evolution was measured for a maximum of 25 min until H_2 concentration curves reached an equilibrium plateau. The rates of H_2 evolution presented in this study are the slopes of H_2 concentration curves before this plateau.

3.5. High-Performance Liquid Chromatography (HPLC)

Acetate concentration was measured with an HPLC apparatus equipped with a HPX-87H anion exchange column (Bio-Rad Laboratories Inc., Hercules, CA, USA) at a temperature of 30 °C, with 5 mM H_2SO_4 as the mobile phase, and a flow rate of 0.6 mL/min.

3.6. SDS-PAGE and Mass Spectrometry

Proteins from 220 mL of *S. ovata* cell-free spent medium were concentrated 375 times with Amicon Ultra-15 centrifugal filter devices with a nominal molecular weight limit of 3 kDa (Merck Millipore, Hellerup, Denmark). Protein concentration was measured with a Coomassie Plus Assay kit (ThermoScientific, Hvidovre, Denmark) and 1.5 μg of protein from two cell-free spent medium samples were loaded on a Sodium Dodecyl Sulfate Poly-Acrylamide Gel Electrophoresis (SDS-PAGE, 12.5%). The PageRuler prestained protein ladder (ThermoScientific) was loaded on the same gel to evaluate the molecular weight of protein bands. After protein separation, the SDS-PAGE was stained with the GelCode Blue stain reagent (Life Technologies, Carlsbad, CA, USA). Revealed protein bands were excised and sent to Alphalyse (Odense, Denmark) for protein identification by matrix-assisted laser desorption/ionization-tandem mass spectrometry (MALDI-MS/MS).

3.7. Hydrogenase Activity Assay

Hydrogen-evolving hydrogenase activity in the cell-free spent medium from a S. ovata-driven MES reactor and in 311 sterile fresh medium was measured in triplicate by monitoring the decrease in the absorbance of dithionite-reduced methyl viologen as described previously [52,53]. Briefly, cell-free spent medium or sterile medium were combined with 0.1M HEPES buffer (pH 8.0) and 100 μM reduced methyl viologen in an anoxic rubber-stoppered cuvette. Change in the absorbance at 604 nm was monitored over time at room temperature. The extinction coefficient of methyl viologen at 604 nm is 13.9 mM^{-1} cm^{-1}.

3.8. Energy-Dispersive X-ray Spectroscopy (EDS)

Duplicate electrode samples were air-dried and examined with a Quanta 200 FEG scanning electron microscope (FEI, Hillsboro, OR, USA). EDS data were collected at an accelerating voltage of 20 kV under high vacuum conditions, with 10 mm working distance and 4.0 spot size.

3.9. Equations

The Nernst Equation (1) was used to calculate theoretical cathode potential (E) at which H_2 evolution starts as described in Vincent et al., 2007 [30].

$$E = E^0 + 2.3RT/nF \log\{(aH^+)^2/p(H_2)\} \tag{1}$$

where E^0 is the standard reduction potential for H_2, R is the gas constant, T is the absolute temperature, n is the number of electrons involved (2 e^- for H_2 evolution), F is the Faraday constant, aH^+ is the activity of H^+ and $p(H_2)$ is the H_2 partial pressure.

Henry's law 2 was used to calculate H_2 partial pressure in the MES reactor gas phase.

$$pH_2 = k_H C \tag{2}$$

where k_H is the Henry's law constant for H_2 and C is the concentration of H_2 in solution.

4. Conclusions

This study showed that in a *S. ovata*-driven MES reactor, faster H_2 evolution at higher cathode potential was enabled because the presence of microbial catalyst modified the cathodic solution chemistry. A higher accumulation of H_2 means more reducing power, which should lead to lower requirements for electrical energy input for microbial CO_2 reduction. The observed acceleration of H_2 evolution could be due to the deposition of cobalt and nickel at the surface of the cathode caused by *S. ovata* catalytic activity. However, given the complexity of chemical species in *S. ovata* cell-free filtrate, it is possible that a more complicated synergistic effect is involved in H_2 evolution during MES with *S. ovata*.

Supplementary Materials: The following are available online at http://www.mdpi.com/2073-4344/9/2/166/s1, Figure S1: Experimental setup for H_2 measurement in close proximity to the cathode surface of a MES reactor, Figure S2: H_2 evolution profile over a period of 25 min with fresh sterile medium in the cathode chamber of a MES reactor, Figure S3: H_2 evolution profile over a period of 25 min with *S. ovata* cell suspension in the cathode chamber of a MES reactor, Figure S4: H_2 evolution profile over a period of 25 min with *S. ovata* cell-free spent medium in the cathode chamber of a MES reactor, Figure S5: Evolution of acetate concentration over time in a MES reactor filled with sterile medium or with *S. ovata* cell-free spent medium with a cathode set at a potential of either -600 mV or -900 mV vs. Ag/AgCl, Figure S6: Proteins found in the cell-free spent medium of electrosynthetic *S. ovata*.

Author Contributions: P.-L.T. and T.Z. conceived the project and designed the experiments. N.F. assembled and operated MES reactors. N.F. prepared *S. ovata* cell suspension, cell-free spent medium and measured H_2 evolution with a hydrogen microsensor. Energy-dispersive X-ray spectroscopy, protein identification via SDS-PAGE and hydrogenase activity assay were done by P.-L.T. and T.Z. P.-L.T. and T.Z. wrote the manuscript with feedback from N.F.

Funding: This work was funded by the Novo Nordisk Foundation, the Chinese Thousand Talents Plan Program and Wuhan University of Technology.

Acknowledgments: We thank Daniel Höglund and Dawid Mariusz Lizak for their assistance with sampling and data processing.

Conflicts of Interest: The authors declare no conflict of interest.

References

1. Tremblay, P.-L.; Zhang, T. Electrifying microbes for the production of chemicals. *Front. Microbiol.* **2015**, *6*, 201. [CrossRef] [PubMed]

2. Rabaey, K.; Rozendal, R.A. Microbial electrosynthesis—Revisiting the electrical route for microbial production. *Nat. Rev. Microbiol.* **2010**, *8*, 706–716. [CrossRef] [PubMed]

3. Nevin, K.P.; Woodard, T.L.; Franks, A.E.; Summers, Z.M.; Lovley, D.R. Microbial electrosynthesis: Feeding microbes electricity to convert carbon dioxide and water to multicarbon extracellular organic compounds. *mBio* **2010**, *1*, e00103–e00110. [CrossRef] [PubMed]

4. Lovley, D.R. Electromicrobiology. *Annu. Rev. Microbiol.* **2012**, *66*, 391–409. [CrossRef] [PubMed]

5. Cheng, S.; Xing, D.; Call, D.F.; Logan, B.E. Direct biological conversion of electrical current into methane by electromethanogenesis. *Environ. Sci. Technol.* **2009**, *43*, 3953–3958. [CrossRef] [PubMed]

6. Ganigué, R.; Puig, S.; Batlle-Vilanova, P.; Balaguer, M.D.; Colprim, J. Microbial electrosynthesis of butyrate from carbon dioxide. *Chem. Commun.* **2015**, *51*, 3235–3238. [CrossRef] [PubMed]

7. Bajracharya, S.; Vanbroekhoven, K.; Buisman, C.J.N.; Pant, D.; Strik, D.P.B.T.B. Application of gas diffusion biocathode in microbial electrosynthesis from carbon dioxide. *Environ. Sci. Pollut. Res. Int.* **2016**, *23*, 22292–22308. [CrossRef] [PubMed]

8. Bajracharya, S.; Yuliasni, R.; Vanbroekhoven, K.; Buisman, C.J.N.; Strik, D.P.B.T.B.; Pant, D. Long-term operation of microbial electrosynthesis cell reducing CO_2 to multi-carbon chemicals with a mixed culture avoiding methanogenesis. *Bioelectrochemistry* **2017**, *113*, 26–34. [CrossRef] [PubMed]

9. Zhang, T.; Tremblay, P.-L. Hybrid photosynthesis-powering biocatalysts with solar energy captured by inorganic devices. *Biotechnol. Biofuels* **2017**, *10*, 249. [PubMed]

10. Jourdin, L.; Raes, S.M.T.; Buisman, C.J.N.; Strik, D.P.B.T.B. Critical biofilm growth throughout unmodified carbon felts allows continuous bioelectrochemical chain elongation from CO_2 up to caproate at high current density. *Front. Energy Res.* **2018**, *6*, 7. [CrossRef]

11. Harnisch, F.; Rosa, L.F.M.; Kracke, F.; Virdis, B.; Krömer, J.O. Electrifying white biotechnology: Engineering and economic potential of electricity-driven bio-production. *ChemSusChem* **2015**, *8*, 758–766. [CrossRef] [PubMed]

12. Kracke, F.; Vassilev, I.; Krömer, J.O. Microbial electron transport and energy conservation—The foundation for optimizing bioelectrochemical systems. *Front. Microbiol.* **2015**, *6*, 575. [CrossRef] [PubMed]

13. Gregory, K.B.; Bond, D.R.; Lovley, D.R. Graphite electrodes as electron donors for anaerobic respiration. *Environ. Microbiol.* **2004**, *6*, 596–604. [CrossRef] [PubMed]

14. Tremblay, P.-L.; Angenent, L.T.; Zhang, T. Extracellular electron uptake: Among autotrophs and mediated by surfaces. *Trends Biotechnol.* **2017**, *35*, 360–371. [CrossRef] [PubMed]

15. Lovley, D.R.; Nevin, K.P. Electrobiocommodities: Powering microbial production of fuels and commodity chemicals from carbon dioxide with electricity. *Curr. Opin. Biotechnol.* **2013**, *24*, 385–390. [CrossRef] [PubMed]

16. Yates, M.D.; Eddie, B.J.; Kotloski, N.J.; Lebedev, N.; Malanoski, A.P.; Lin, B.; Strycharz-Glaven, S.M.; Tender, L.M. Toward understanding long-distance extracellular electron transport in an electroautotrophic microbial community. *Energy Environ. Sci.* **2016**, *9*, 3544–3558. [CrossRef]

17. Aulenta, F.; Reale, P.; Catervi, A.; Panero, S.; Majone, M. Kinetics of trichloroethene dechlorination and methane formation by a mixed anaerobic culture in a bio-electrochemical system. *Electrochim. Acta* **2008**, *53*, 5300–5305. [CrossRef]

18. Summers, Z.M.; Gralnick, J.A.; Bond, D.R. Cultivation of an obligate Fe(II)-oxidizing lithoautotrophic bacterium using electrodes. *mBio* **2013**, *4*, e00420. [CrossRef]

19. Deutzmann, J.S.; Sahin, M.; Spormann, A.M. Extracellular enzymes facilitate electron uptake in biocorrosion and bioelectrosynthesis. *mBio* **2015**, *6*, e00496. [CrossRef]

20. Jourdin, L.; Lu, Y.; Flexer, V.; Keller, J.; Freguia, S. Biologically induced hydrogen production drives high rate/high efficiency microbial electrosynthesis of acetate from carbon dioxide. *ChemElectroChem* **2016**, *3*, 581–591. [CrossRef]

21. LaBelle, E.V.; Marshall, C.W.; Gilbert, J.A.; May, H.D. Influence of acidic pH on hydrogen and acetate production by an electrosynthetic microbiome. *PLoS ONE* **2014**, *9*, e109935. [CrossRef] [PubMed]

22. May, H.D.; Evans, P.J.; LaBelle, E.V. The bioelectrosynthesis of acetate. *Curr. Opin. Biotechnol.* **2016**, *42*, 225–233. [CrossRef] [PubMed]

23. Kundu, A.; Sahu, J.N.; Redzwan, G.; Hashim, M.A. An overview of cathode material and catalysts suitable for generating hydrogen in microbial electrolysis cell. *Int. J. Hydrogen Energy* **2013**, *38*, 1745–1757. [CrossRef]

24. Call, D.; Logan, B.E. Hydrogen production in a single chamber microbial electrolysis cell lacking a membrane. *Environ. Sci. Technol.* **2008**, *42*, 3401–3406. [CrossRef] [PubMed]

25. Cheng, S.; Logan, B.E. High hydrogen production rate of microbial electrolysis cell (MEC) with reduced electrode spacing. *Bioresour. Technol.* **2011**, *102*, 3571–3574. [CrossRef] [PubMed]

26. Hu, H.; Fan, Y.; Liu, H. Hydrogen production using single-chamber membrane-free microbial electrolysis cells. *Water Res.* **2008**, *42*, 4172–4178. [CrossRef] [PubMed]

27. Lee, H.-S.; Salerno, M.B.; Rittmann, B.E. Thermodynamic evaluation on H_2 production in glucose fermentation. *Environ. Sci. Technol.* **2008**, *42*, 2401–2407. [CrossRef] [PubMed]

28. Lee, H.-S.; Rittmann, B.E. Characterization of energy losses in an upflow single-chamber microbial electrolysis cell. *Int. J. Hydrogen Energy* **2010**, *35*, 920–927. [CrossRef]

29. Selembo, P.A.; Merrill, M.D.; Logan, B.E. The use of stainless steel and nickel alloys as low-cost cathodes in microbial electrolysis cells. *J. Power Sources* **2009**, *190*, 271–278. [CrossRef]

30. Vincent, K.A.; Parkin, A.; Armstrong, F.A. Investigating and exploiting the electrocatalytic properties of hydrogenases. *Chem. Rev.* **2007**, *107*, 4366–4413. [CrossRef] [PubMed]

31. Jeremiasse, A.W.; Hamelers, H.V.M.; Kleijn, J.M.; Buisman, C.J.N. Use of biocompatible buffers to reduce the concentration overpotential for hydrogen evolution. *Environ. Sci. Technol.* **2009**, *43*, 6882–6887. [CrossRef]

32. Liang, D.; Liu, Y.; Peng, S.; Lan, F.; Lu, S.; Xiang, Y. Effects of bicarbonate and cathode potential on hydrogen production in a biocathode electrolysis cell. *Front. Environ. Sci. Eng.* **2014**, *8*, 624–630. [CrossRef]

33. De Silva Muñoz, L.; Bergel, A.; Féron, D.; Basséguy, R. Hydrogen production by electrolysis of a phosphate solution on a stainless steel cathode. *Int. J. Hydrogen Energy* **2010**, *35*, 8561–8568. [CrossRef]

34. Merrill, M.D.; Logan, B.E. Electrolyte effects on hydrogen evolution and solution resistance in microbial electrolysis cells. *J. Power Sources* **2009**, *191*, 203–208. [CrossRef]

35. Daniele, S.; Lavagnini, I.; Baldo, M.A.; Magno, F. Steady state voltammetry at microelectrodes for the hydrogen evolution from strong and weak acids under pseudo-first and second order kinetic conditions. *J. Electroanal. Chem.* **1996**, *404*, 105–111. [CrossRef]

36. Rosenbaum, M.; Aulenta, F.; Villano, M.; Angenent, L.T. Cathodes as electron donors for microbial metabolism: Which extracellular electron transfer mechanisms are involved? *Bioresour. Technol.* **2011**, *102*, 324–333. [CrossRef] [PubMed]

37. Yates, M.D.; Siegert, M.; Logan, B.E. Hydrogen evolution catalyzed by viable and non-viable cells on biocathodes. *Int. J. Hydrogen Energy* **2014**, *39*, 16841–16851. [CrossRef]

38. Tremblay, P.-L.; Höglund, D.; Koza, A.; Bonde, I.; Zhang, T. Adaptation of the autotrophic acetogen *Sporomusa ovata* to methanol accelerates the conversion of CO_2 to organic products. *Sci. Rep.* **2015**, *5*, 16168. [CrossRef] [PubMed]

39. Aryal, N.; Tremblay, P.-L.; Xu, M.; Daugaard, A.E.; Zhang, T. Highly conductive poly(3,4-ethylenedioxythiophene) polystyrene sulfonate polymer coated cathode for the microbial electrosynthesis of acetate from carbon dioxide. *Front. Energy Res.* **2018**, *6*, 72. [CrossRef]

40. Aryal, N.; Tremblay, P.-L.; Lizak, D.M.; Zhang, T. Performance of different *Sporomusa* species for the microbial electrosynthesis of acetate from carbon dioxide. *Bioresour. Technol.* **2017**, *233*, 184–190. [CrossRef]

41. Zhang, T.; Nie, H.; Bain, T.S.; Lu, H.; Cui, M.; Snoeyenbos-West, O.L.; Franks, A.E.; Nevin, K.P.; Russell, T.P.; Lovley, D.R. Improved cathode materials for microbial electrosynthesis. *Energy Environ. Sci.* **2013**, *6*, 217–224. [CrossRef]

42. Nie, H.; Zhang, T.; Cui, M.; Lu, H.; Lovley, D.R.; Russell, T.P. Improved cathode for high efficient microbial-catalyzed reduction in microbial electrosynthesis cells. *Phys. Chem. Chem. Phys.* **2013**, *15*, 14290–14294. [CrossRef] [PubMed]

43. Ammam, F.; Tremblay, P.-L.; Lizak, D.M.; Zhang, T. Effect of tungstate on acetate and ethanol production by the electrosynthetic bacterium *Sporomusa ovata*. *Biotechnol. Biofuels* **2016**, *9*, 163. [PubMed]

44. Aryal, N.; Halder, A.; Tremblay, P.-L.; Chi, Q.; Zhang, T. Enhanced microbial electrosynthesis with three-dimensional graphene functionalized cathodes fabricated via solvothermal synthesis. *Electrochim. Acta* **2016**, *217*, 117–122. [CrossRef]

45. Aryal, N.; Halder, A.; Zhang, M.; Whelan, P.R.; Tremblay, P.-L.; Chi, Q.; Zhang, T. Freestanding and flexible graphene papers as bioelectrochemical cathode for selective and efficient CO_2 conversion. *Sci. Rep.* **2017**, *7*, 9107. [CrossRef] [PubMed]

46. Chen, L.; Tremblay, P.-L.; Mohanty, S.; Xu, K.; Zhang, T. Electrosynthesis of acetate from CO_2 by a highly structured biofilm assembled with reduced graphene oxide–tetraethylene pentamine. *J. Mater. Chem. A* **2016**, *4*, 8395–8401. [CrossRef]

47. Ragsdale, S.W.; Pierce, E. Acetogenesis and the Wood–Ljungdahl pathway of CO_2 fixation. *Biochim. Biophys. Acta* **2008**, *1784*, 1873–1898. [CrossRef]

48. Ragsdale, S.W. Nickel-based enzyme systems. *J. Biol. Chem.* **2009**, *284*, 18571–18575. [CrossRef]

49. Ragsdale, S.W. Enzymology of the Wood–Ljungdahl pathway of acetogenesis. *Ann. N. Y. Acad. Sci.* **2008**, *1125*, 129–136. [CrossRef]

50. Yu, N.Y.; Wagner, J.R.; Laird, M.R.; Melli, G.; Rey, S.; Lo, R.; Dao, P.; Sahinalp, S.C.; Ester, M.; Foster, L.J.; et al. PSORTb 3.0: Improved protein subcellular localization prediction with refined localization subcategories and predictive capabilities for all prokaryotes. *Bioinformatics* **2010**, *26*, 1608–1615. [CrossRef]

51. Möller, B.; Oßmer, R.; Howard, B.H.; Gottschalk, G.; Hippe, H. *Sporomusa*, a new genus of gram-negative anaerobic bacteria including *Sporomusa sphaeroides* spec. nov. and *Sporomusa ovata* spec. nov. *Arch. Microbiol.* **1984**, *139*, 388–396. [CrossRef]

52. Nakashimada, Y.; Rachman, M.A.; Kakizono, T.; Nishio, N. Hydrogen production of *Enterobacter aerogenes* altered by extracellular and intracellular redox states. *Int. J. Hydrogen Energy* **2002**, *27*, 1399–1405. [CrossRef]

53. Fernández, V.M.; Gutiérrez, C.; Ballesteros, A. Determination of hydrogenase activity using an anaerobic spectrophotometric device. *Anal. Biochem.* **1982**, *120*, 85–90. [CrossRef]

MDPI

St. Alban-Anlage 66

4052 Basel

Switzerland

Tel. +41 61 683 77 34

Fax +41 61 302 89 18

www.mdpi.com

Catalysts Editorial Office

E-mail: catalysts@mdpi.com

www.mdpi.com/journal/catalysts